FINEST LONDON PORTLAND CEMENT
ZENITH
THE HIGHEST
QUALITY
MADE IN
ENGLAND

EASTWOODS, LIMITED
WELLINGTON
BRAND
LONDON
PORTLAND CEMENT

LONDON
ENGLAND

FRANCIS & Co
THE NINE ELMS BRAND REGISTERED
PORTLAND CEMENT
MANUFACTURED IN
ENGLAND
LONDON

MANUFACTURED BY JOHN BAZLEY WHITE & BROTHERS LIMITED, LONDON, ENGLAND.
SEMPER IDEM

THE WOULDHAM CEMENT COMPANY LIMITED
LONDON
UNION JACK
GUARANTEED PURE PORTLAND
WORKS: LION WORKS
WEST THURROCK, ESSEX
ENGLAND.

THE ASSOCIATED PORTLAND CEMENT MANUFACTURERS LIMITED
PORTLAND
CEMENT
PRIZE
MEDAL
Antwerp
Exhibition
1885
BRITANNIA BRAND
LONDON

THE BRITISH PORTLAND CEMENT MANUFACTURERS LIMITED
REGD
TRADE MARK
SUN
BRAND
PORTLAND
CEMENT
MADE IN ENGLAND
LONDON

THE ASSOCIATED PORTLAND CEMENT MANUFACTURERS LIMITED
LONDON
TRADE MARK
REGD
PORTLAND CEMENT
LONDON

PORTLAND
KNIGHT, BEVAN & STURGE
LONDON
CEMENT

PORTLAND CEMENT
ROBINS & Co
THREE KILNS BRAND
LONDON & NORTHFLEET, KENT.

BEST LONDON PORTLAND CEMENT
TRADE MARK
REGISTERED
BROOKS, SHOOBRIDGE & Co

IMPERIAL PORTLAND CEMENT Co
REGISTERED
TRADE MARK
FLAG BRAND
NORTHFLEET, ENGLAND

THE ASSOCIATED PORTLAND CEMENT MANUFACTURERS LIMITED
HILTON'S
PORTLAND CEMENT
LONDON
GREAT BRITAIN

THE WOULDHAM CEMENT COMPANY LIMITED
LONDON
DELTA
BRAND
D
LION WORKS:
WEST THURROCK, ESSEX,
ENGLAND.

CEMENT, MUD AND 'MUDDIES'

CEMENT, MUD AND 'MUDDIES'

By
FRANK G. WILLMOTT

Photography by
Alan Cordell

MERESBOROUGH BOOKS

1977

Further copies of this book may be obtained through your local bookshop or from F.G. Willmott, 7 Wakeley Road, Rainham, Kent. Price £2.95 plus 35p postage and packaging.

Trade orders to Meresborough Books, 14 London Road, Rainham, Kent.

By the same author:

Bricks and 'Brickies'

The story of Eastwoods the Brickmakers and their fleet of sailing barges. Now reprinted at £2.50 plus 30p postage and packaging.

ISBN 0 905270 02 9

Printed in Great Britain by
W & J Mackay Limited, Chatham
by photo-litho.

CONTENTS

PART IV: BARGES AND FREIGHTS

PART V: THE CLAY WORK

PART VI: BARGE HISTORIES

LIST OF ILLUSTRATIONS

PHOTOGRAPHER'S ACKNOWLEDGEMENT

The photographs in this publication were obtained in two ways. Some were from my own collection of negatives and others were obtained by borrowing original photographs and making negatives and prints from them.

Thus I would like to thank the people who are listed below for their help in lending old photographs; Thames Barge Sailing Club. T. Redshaw. R. Stimson Jnr and H. Oliver Hill.

A. CORDELL

ACKNOWLEDGEMENTS

I am most grateful

I am most grateful for the help and useful information given to me by Captains F. Farrington, C. White, F. Whitnell, J. Jewsberry and V. Wadhams.

Mr H. Wickenden, S. Simmonds, J. Bath, C. Croucher, B. London, B. Rayner, P. Packer, F. (Skip) Eldridge, N. Cheeseman, B. Squires and L. Hill.

The British Aerofilms Ltd., Maidstone Museum, and Gillingham Public Library.

To Mr. Joseph Hines who checked the list of barges.

To an article by Mr F.F. Woodruff, published in an early edition of "Sea Breezes".

To Mr J.M. Preston, B.A., D.P.A., College lecturer of history.

I am especially grateful to Mr Alec J. Kean, O.B.E., who made much material available to me, and offered help on further occasions.

To Mrs Linda Hampton for ridding the manuscript of my worst grammatical faults.

To Skipper Bob Childs for his final reading of the manuscript and additions to historical notes.

The half-tones and sketches were kindly drawn by my very good friend Bernard Kent.

F.G.W.

Part I: The Cement Industry Develops

This is my second book on the river-borne industries of Kent.

It contains a history of the vast use of river mud or London clay that has been and is still being used as an additive to the chalk in the manufacture of cement in North Kent.

So although this book is about mud and muddies, it is inevitable that I briefly mention the histories of some of the more well known cement factories that were thriving at the turn of the century, together with some of the trades that were allied to it, like the digging of the London clay, coopering and in particular the owning, building and the handling of the Spritsail Sailing Barge.

Cement is a cheap heavy commodity, manufactured from bulk low value material such as chalk and river mud and using considerable amounts of fuel, namely coal.

In the early days, most of the chalk was dug from the bluffs close to the river, while the clay, coal or coke was brought in by barge.

The chalk found near the river Thames and Medway and the mud or clay from the rivers was first used by the Romans, and cement making has progressed from that time. So then with an abundant supply of raw materials found in the chalk formations of the North Downs, together with the clay found near the rivers, we have the following table of a typical analysis of these materials mixed on approximately five of chalk to two of clay basis.

	Chalk	*River Mud*
Silica	1.52	65.02
Alumina)	0.85	19.74
Ferric Oxide)		7.22
Carmonate of Lime	96.45	1.52
Loss, etc. (Organic Matter)	1.18	6.50

EARLY YEARS

The cement industry in North Kent put down its roots in the days of lime burning.

Large quantities of the lime used in the advanced farming counties of Essex and Suffolk at the end of the eighteenth century came from the works on the upper Medway.

The area was also a major supplier of building lime and whiting to the London builders merchants, lime from Halling reputedly being used in the construction of the Waterloo and new London bridges.

Lime was much more difficult to store and transport than cement, and so was usually prepared where it was to be used. As the means of transport improved, i.e. many chalk barges were working the two rivers, chalk could be taken longer distances.

In cement making, however, no advances had been made since Roman times, until 1796, when one Joseph Parker of Northfleet, patented his "Roman Cement" made by burning nodules of argillaceous chalk found in the London clay of the locality and around the Kentish coast.

Joseph Aspdin, a Leeds man, is wrongly thought to have invented 'Portland' cement. In fact 'Portland' cement in a modern form was first made at the John Bazley Whites works at Swanscombe about 1845. Its discoverer was a Mr I.C. Johnson, the Swanscombe Works Manager, although Aspdin was first to patent Portland cement in this name as early as 1824 since it resembled Portland stone. The Patent No was 5022. Two years previously, James Frost had patented his 'British' cement, using a 2–1 chalk to Medway clay; he did start large scale production at Swanscombe in 1825.

In 1826, Sir C.W. Pasley made cement with properties similar to that of Aspdins by burning a mixture of chalk and river mud.

THAMES WORKS

In 1850 there were still only four Portland cement works on the Thames, but after 1851 the number of works on the Thames and Medway rapidly increased.

Aspdin had sited a factory in the grounds of his own home at Northfleet. His specifications on the patent were insufficiently detailed so others were unable to use his methods. So it was that various others like Johnson, Bevans and Whites built works beside the Thames and produced Portland cement. But Portland cement was not immediately an overwhelming success due to the often poor quality of the product which resulted from the rule of thumb methods used in mixing the ingredients and firing them in the kiln. However it had become an acceptable material in wide demand by about 1870, and this has not declined up to this time. This was to a large extent as a result of improvements in testing due to the work of a Mr J. Grant of the Metropolitan Board of Works, who required a reliable cement for his early drainage schemes.

MEDWAY WORKS

The Medway Valley used to be of no mean importance either. On the first of May 1851, I.C. Johnson opened the first Portland Cement Works on the Medway at Frindsbury. The site he used was an abandoned oil seed crushing mill adjacent to the river where there existed an old Boulton & Watt steam engine to drive the machinery.

The position of the Ness at Frindsbury was ideal, with abundant supplies of chalk from the spur behind, and almost unlimited quantities of the other main ingredients, in particular the blue river mud and the large amounts of coal that were being shipped to Rochester.

Water transport, which was relatively cheap, was another facility offered by the area and this, in turn, helped the barge builders like Gill, Curel and Highams that were established in this area.

The first cement factory at Frindsbury began producing around fifty tons a week. The process of expansion at Frindsbury was presided over by William Tingey who acquired the leases ot the Chatham Ness area from the Bridge Wardens and the Church Commissioners of Rochester.

Over the next forty years there were no less than seven works built at Frindsbury. This was the biggest concentration of works ever, the cement companies being Bridge, Pheonix, Globe, Quarry, Crown, Beehive, and Beaver, all crowding the water's edge. When this complex was completed in 1900, it consisted of 152 chamber kilns of various types with about thirty tall brick chimneys. The factories employed around 800 men and boys, and had an output of some 4,000 tons per week.

Further round the Ness was yet another small cement works built and worked by the Formby Brothers, James and Charles; this works produced about six hundred tons a week. No less than twenty-one works have been set up in the Frindsbury area since 1851 when the first works was opened, although several smaller mills had been in operation in the area previously.

The first cement factory was built in 1860 at Wouldham in the centre of the village; this was known as The Wouldham Cement Works. Later the Peters Brothers opened works at Burham and was known as the Burham Brick Cement and Lime Company. In 1880 however, Burham could boast as being one of the largest cement factories in the world at that time.

As the metropolis of London continued to expand, together with the exportation of British cement abroad, there came a bigger demand

The Wickham Works of Martin Earle & Co. looking north.

The Wickham Works of Martin Earle & Co. looking south.

for more cement. With improved production methods and more works opening up, the Thames and Medway had a world monopoly of production, selling their products not only in Europe, but also in the U.S.A., the Argentine, Australia, South Africa and India.

The cement for the Aswan Low Dam came from Frindsbury and the cement for repairing the damage caused by the San Francisco earthquake of 1909 came from Gillingham cement works.

By 1900 there were twenty-six works from Burham downstream to

Rainham and in the Medway-Thames region, there were up to forty cement works operating.

The biggest single factory on the Medway at the turn of the century was the Wickham Works of Martin Earle & Co. The works were begun in 1881 on a small scale by John Adams of Strood, and after changing hands several times, and the bankruptcy of the Reliance Portland Cement Company in 1893, came into the hands of J.B. Martin, a local Veterninary Surgeon. He, in partnership with E.J.V. Earle, expanded the works until in 1899, they produced 3,000 tons of cement a week with a labour force of something like 2,000 hands. The employment of such large numbers of men and boys was necessary due to the amount of manual handling of raw and finished materials in such operations as charging the kilns and quarrying, with nothing more than a pick and shovel. Even so the heavy and sometimes dangerous work found a good supply of labour from amongst the poorly paid agricultural workers of the county.

The Wouldham Cement Factory closed in 1903 and the Burham Factory closed in 1925.

The works at Holborough was built in 1926 on a completely new site, using a new cement-making technique.

This works was taken over from the Red Triangle Group by the A.P.C.M. in 1931.

Lee's of Halling shut down as a direct result of its competition.

Profound technological changes were also about to take place from the 1890's onwards, with the introduction of ball and roller grinding mills and the revolutionary rotary kiln. Apart from the many difficulties already suffered through competition, many of the older works had insufficient capital to modernise sufficiently to remain competitive.

Some works were modernised with the introduction of rotary kilns at Gillingham, Wouldham Hall, and a battery of four at Burham.

So cement expanded production rapidly, with the new plant coming into operation leading to a situation of intense competition by the end of the 1890-1900's. Cement can be produced anywhere that the right combination of materials occurs. Cement works set up in other parts of Britain immediately gained virtual monopolies of local markets due to advantages in freight costs.

Cement produced on the river Medway had to be taken by barge to the London Docks for shipment, adding expensive handling costs.

Even for the London market the cement had to be bagged or casked and taken by barge a longer distance round the Isle of Grain than cement from the Thames, and this made it marginally more expensive.

Rochester Bridge also was a considerable obstacle to the shipment of cement from the upper Medway, and added to the geographical isolation of the works on the right bank at Borstal, Wouldham and Burham. Works in this area were at a particular disadvantage, due to the lack of good road access and rail communications.

The works of Peters Brothers at Burham shipped cement across the Medway to the goods yard at Halling for some considerable while in order to keep going, but ultimately the result was the wholesale closure of works, including six of the Frindsbury Group, the two Burham works and the Wouldham works.

The answer to the problems of the cement industry appeared to be through some form of regulation of production and prices. This came about in a partial form through two amalgamations in the early part of the century.

Firstly 1900, there was the formation of the A.P.C.M., secondly in 1912 an unsuccessful attempt to merge the remaining independent firms in the Red Triangle Group or later the B.P.C.M. The more prominent firms here were the Holborough works, Smeed Deans of Murston and Goldsmiths of Rainham.

The aim of the mergers was to concentrate production on the most modern and efficient units, or those capable of expansion and to streamline the marketing operations. The inter-war problems of over production, although partially overcome by quota arrangements, led to the disappearance of all but two of the pre-1914 works, the Wickham works, absorbed into the B.P.C.M., in the 1920's and the Crown works of A.P.C.M. which closed in 1956 when all the chalk that could be excavated in the area became exhausted. This, the last works at Frindsbury, was finally demolished in 1973.

Two other works to survive were the Holborough works, established in 1926 and later acquired by A.P.C.M. Ltd. and the Halling works of the Rugby Portland Cement Co. built in 1936.

All these works are sited on the west bank of the River Medway.

Another factor was that the sailing barge building Industry was dying just before 1914–18 war.

The problem for the Medway was that unlike the works at Swanscombe and Northfleet where vessels of up to 10,000 tons could come alongisde to load, 500 ton vessels were the limit at Frindsbury.

When the new Northfleet works comes into full production, it will be capable of producing 4 million tons a year. The deep water facilities there are capable of docking large bulk carriers of up to 60,000 tons. The building of this large new plant at Northfleet resulted in several

An ariel photo of the Gillingham Portland Cement Works. To the left are the remains of the John Bazley White's works. In the middle are the remains of the Gillingham Whiting Works owned by the Gillingham Syndicate. The rigged barge in the centre of the photo is the *Daisy Little* being unloaded at the Clay berth. The lighter being unloaded under the crane is the old "swimmie" *Argossy* now renamed *Bulldog*.
The other two craft moored off singularly are the "swimmies" *Kelley* and *Prudent*.

other works closing down, namely Murston, Cliffe and Swanscombe.

However the cement industry has left its permanent mark in the form of the huge quarries scarring the landscape of the Thames and Medway Valley as a monument to past glories.

GRAVESEND – ROCHESTER CANAL

The construction of this 6¾ mile canal was financed by a group which was called the Thames and Medway Canal Company in 1824.

The purpose of the canal was to eliminate the long passage from Rochester down river round the Isle-of-Grain and up Sea Reach to Gravesend by the many chalk, lime and cement barges which were working to London from the upper reaches of the Medway.

The canal was a unique project at the time, due to the final two and a half miles leading to Strood having to be tunnelled through the solid chalk of Higham Hills. The canal was never a commercial success, because at Strood where the canal ran into the Medway, a massive steam engine had to be used to pump water into the canal at ebb tides.

Although the Government at the time looked favourably on the scheme because it protected many barges from the deprivations of privateers in the Thames estuary, for any nautical use it was declared a 'white elephant' and was sold to the Railway Company.

BARGES IN THE CLAY WORK

Generally a cement firm would have about four or five barges working from the clay beds to the factory but sometimes when the clay was getting a bit low they would put on an extra barge.

Barges in the clay work were usually of a large capacity and generally had their holds divided into three compartments by fitting a series of baffle boards across the holds; this was to stop the wet clay from slumping from one side while under way.

These clay barges had very good gear and had experienced skippers, this was so a barge could keep going in all conditions to keep up the supplies of clay to the works. If the weather was rough, squally or foggy, a clay barge would have to get underway somehow because if the works ran out of clay, they would literally grind to a halt.

Some of the barges in the clay work were the old 'tear-outs' that only did short trips. Many craft finished their working lives like this or this work usually finished them. Several barges would go into a mudhole on the flood and would moor up in a spot marked with two poles or sticks set in by the muddies who were to load her. Sometime before the tide had receded, a chain was passed under the barge in case she got stuck down with suction as many did, for when a barge was

DISCHARGING OVERSIDE
ON THE HARD.

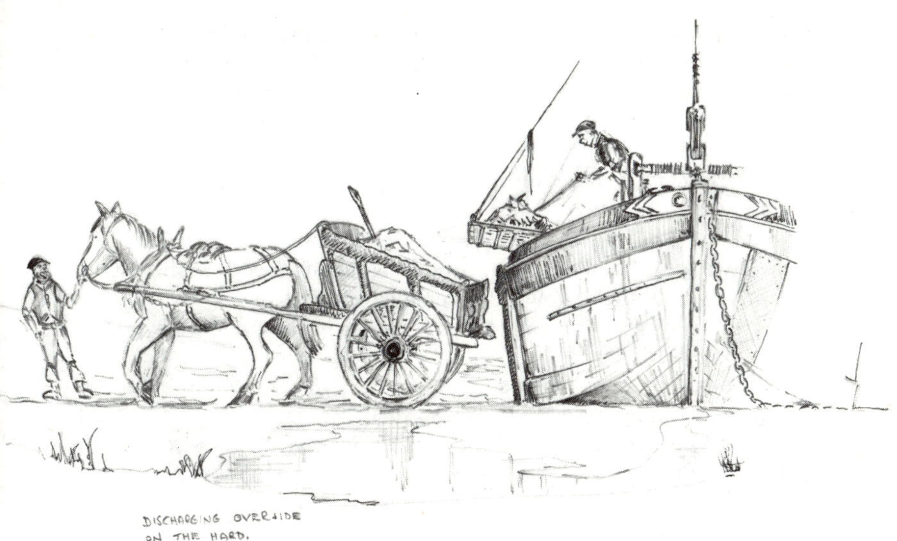

A barge unloading clay by hand.

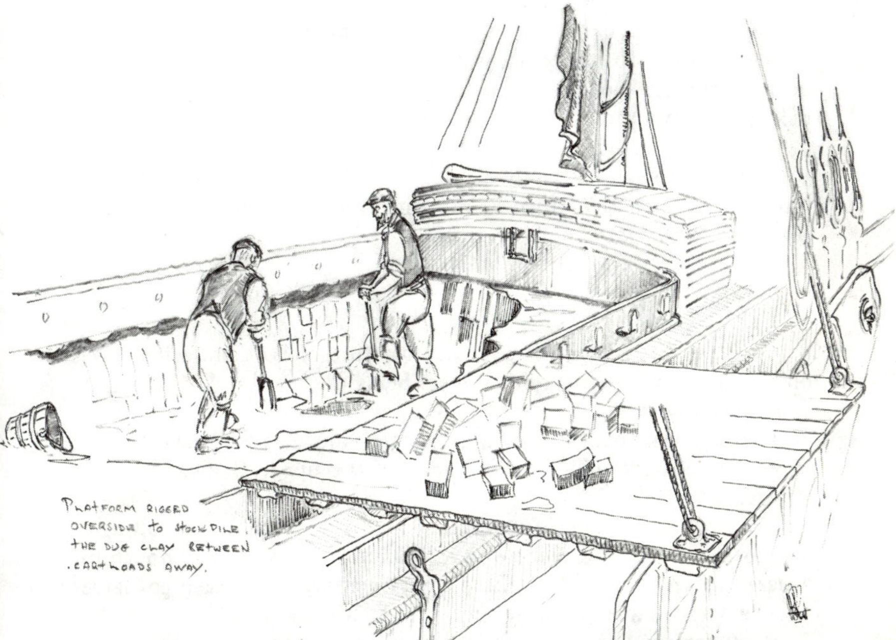
PLATFORM RIGGED
OVERSIDE to STOCKPILE
THE DUG CLAY BETWEEN
CARTLOADS AWAY.

Unloading clay by hand.

loaded the weight of clay and the suction in the mud prevented her lifting on the next tide. This could make a difference of an hour or so in a barge getting away. Usually the mate or skipper would wrap lime cloths of canvas round the mast case and round the dead eyes to give the working parts protection agains the flying clay; a gang of muddies could become very upset at times, especially when they did not get any tea brewed on board.

Sometimes a barge only needed a few inches of water to slip off of a mud berth and this was helped by setting down a kedge anchor with a bridle wire from the brail winch or crab winch aft.

As the clay was dug and loaded, terraces in the bank were formed about three feet wide.

On some occasions when a barge was nearly loaded and the tide was making, the muddies would get out their big old shovels and chuck great big lumps of clay onto the barge's deck weighing nearly half a hundreweight, for the mate of the barge to shovel down the hold, otherwise a barge wouldn't get away on that tide.

When the clay barge had arrived at the works, they usually had a special clay unloading berth. This was a 'low' berth or 'drawdock' so that the clay could be unloaded by crane into carts or railway trucks.

In the early days when clay was unloaded by hand, the unloading would start by digging a hole in the clay, then tipping a bucket of water in it which would make the digging considerably easier. Another thing the gang would do, particularly if there was a shortage of carts, was to set up a canterlevered platform over the barge's side, the overhang being held up by two ropes to a tackle off the spreet, so while the cart was away the men could carry on unloading onto the platform then when the cart returned they could release the platform and shoot the clay into the cart for a quick turn round. With everything done at piecework, time was always at a premium.

CLAYHOLES

From the early 1850's when the manufacture of Portland cement was on the increase, large clayholes were being dug on the saltings on the River Medway.

The clay was loaded into barges by the 'muddies' and taken to the expanding cement mills on the banks of the upper Medway and the River Thames.

The leader of a gang of muddies would pick a spot on the salting to dig. As the digging proceeded, they would form a basin big enough for a barge to get into and as the digging increased, the basin got larger until up to six barges could get into it to be loaded.

Many cement firms had special barges built for the clay traffic. They were of a large capacity, had wide decks and were shallow sided so they could get in and out of a clayhole with the minimum amount of water.

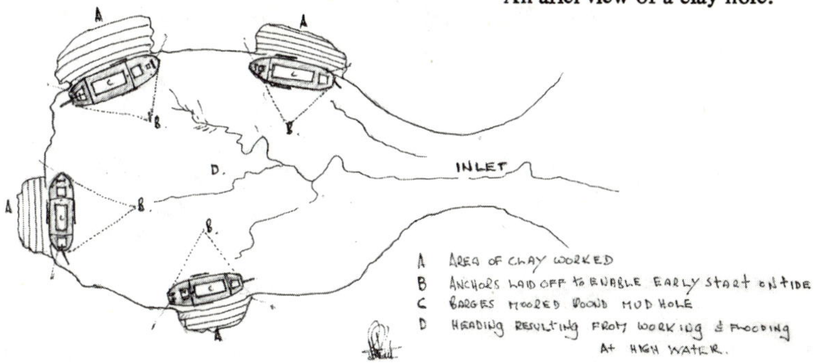

An ariel view of a clay hole.

At the turn of the century one could see up to a hundred craft working the clay away to the works, but after the formation of the 'combine' and the decision to close many of the old works with obsolete and out of date machinery, the number of barges began to dwindle.

By 1910 an Admiralty report stated that they were becoming very concerned about the continual digging of clay from saltings in the Medway estuary, some of the clay workings having reached enormous proportions. The report stated that after a survey of the river it revealed that the large clay workings were disrupting the course of the Channel, and that there were great changes in the river bed, especially in the area of Bishops Marsh.

Between the years 1898–1908, an estimated area of between 15–17 acres or 280,000 tons of clay was dug from various marsh saltings of Barksore, Oakham, Burntwick, Slayhills Millford Hope Marshes.

From the claybeds of West Hoo Creek, between the years 1888–1910 an area of 34–37 acres of 557,000 tons of clay had been dug.

In addition to these figures, several thousands of tons were dug from various points without permission. However with permission from the Medway Conservancy Board, an estimated 3,704,000 tons of clay had been dug, and has caused the enlargement of the tide by something like three square miles.

Around the 1920's the A.P.C.M. had a small fleet of barges doing the mudwork: several of these had steel tanks built in their holds which gave them a few more years of working life.

An early map showing the Thames and Medway Estuaries before clay digging began in a big way. For an idea of the amount of clay dug away one should compare this map with an up to date O.S. Map.

Thames Haven Branch Railway
Sea Reach
The Lower Hope
Cliffe Marsh
Cooling Marsh
Halstow Marsh
Higham Marsh
Cliffe
Cliffe Creek
Higham Creek
Chalk Quarry
Decoy Farm
Canal
Higham
Hoo Common
Hoo
Cockham Wood
Cockham wood Reach
Hoo Salt Marsh
Short Reach
Gillingham Reach
Brickfield
Brompton
Gillingham
ROCHESTER

RIVER THAMES
River Middle Buoy
Leigh Middle Buoy
Boundary of the Jurisdiction
of the Lord Mayor of London
Conservator of the River Thames.
Yantlet Flats
Jenkin Buoy
Jenkin Sand
Slough
The Clinker
Coastguard Sta
Allhallows
Marsh
Yantlet Creek
Oldharbour Barn
Grain Marsh
Dagnam
Windhill
Allhallows
Belle
Yantlet Creek
Sea Wall
Stoke Marsh
Lower Stoke
Middle Stoke
Upper Stoke
Stoke
Malmaines Hall
Stoke Farm
Stoke Salting
Grain Bridge
THE ISLE OF GRAIN
Home Farm
Pilot marks
Sea Wall
Colemouth Creek
Cockleshell Hard
Coast Guard Sta
Black stakes
Port Victoria
Terminus
Saltpan Reach
RIVER MEDWAY
Hoo Marsh
Stede Ouze
Long Reach
Kethole Reach
Bishops Salt Marsh
Bishops Ouze
South Yantlet Creek
Half Acre
Nor Marsh
Saltings
Bartlett Creek
Stangate Spit
Stangate Creek
Deadmans Island
Slayhill Marsh
Greenborough Marsh
Chitney Marsh
Lower Chitney
Upper Chitney
New Chitney
Long Reach
Ham Ouze
Ham Green Marsh
Barksore Marsh
Iwade Marsh

An artist's impression of the floating grab crane and slurry-tank at the Cliffe clay basins.

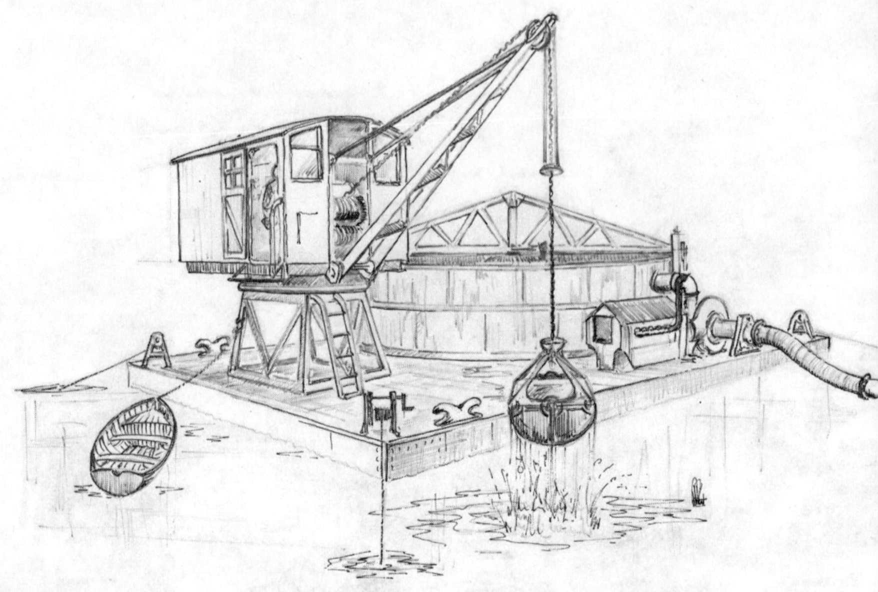

The barges were loaded by Brices grab on Kingsnorth Marshes and sailed round to the Thames works at Northfleet and Swanscombe to be unloaded by grab, then sailed back round again, the crews spending nearly every night in Sea Reach coming and going.

This loading by grab soon knocked the best out of these barges: a grabfull of clay weighed about a ton and this dropped from a height of twelve or fourteen feet say a hundred times a load, gives one a rough idea of what they had to put up with. Not only that, but being continually loaded with mud made them wet and heavy and the only chance they had was if it blew hard, then they would show a good turn of speed. These grab cranes could load a hundred and thirty ton barge in about two hours.

As the wooden sailing barges became old and unseaworthy many sank or would not lift after loading. These craft were left in the mud or burnt or built into sea walls. But by this time the A.P.C.M. were now using some big Lighters which they had built which could carry two hundred tons each in their three separate holds.

Several of these Lighters at a time were towed about by one of the A.P.C.M's tugs *Blue Circle, Colourmist, Colour Crete, Panther* and *Lion*. The Lighters were named after colours *Purple, Mauve & Crimson*, trees – *Oak, Elm, Sycamore* and *Willow*, flowers – *Lupin, Iris, Nasturtium* and *Carnation*.

The lighters were towed round to a steam grab in Colemouth Creek to be loaded.

So as not to infringe any rules of the Freeman of Thames and Medway Watermen, the lighters were towed by the A.P.C.M. tugs to Garrison Point where they were picked up by another tug that worked on the Medway and taken to be loaded and then the same order was kept when returning. This system did not go on for long as the A.P.C.M. began systematically digging clay in a big way on the Cliffe Marshes.

As the manufacturing of cement became more economically minded, the A.P.C.M. began digging clay in large quantities from the Thames. They had over the years dug clay from Higham Bight and Cliffe Creek, much of it going to the nearby works of Francis & Co., and the *Alpha* where hand loading went on as late as 1928. This was loaded into the side tipping *Jubilee* trucks, pulled up to the works by cable.

In the 1930's the A.P.C.M. built a special type of vessel which could dig the clay and turn it into slurry in a washmill that was built on the vessel. An electrically operated crane was also built, this being on a high pedestal above the washmill. The slurry flowed over a weir into a tank where it was pumped through a pipe up to the A.P.C.M. jetty about a mile away. As the clay was grabbed out, a series of clay basins were formed; these were at a depth of fifteen feet.

This vessel was bombed and sunk during the Second World War, but was raised and got back into working order. It was, however, finally broken up in the 1960's.

When the clay reached the jetty it was pumped into one of A.P.C.M.'s three special clay carrying vessels which were now operating at this time. These vessels could carry a freight of 600 tons of clay; their names were *Claycarrier I, Claycarrier II* and *Clay Transporter*. They were based at the 'Kent' works at Swanscombe.

So then by this method the A.P.C.M. were always able to keep up the supplies of clay to the Thames works that were in production at that time. This method proved to be very satisfactory and economical way of supplying clay.

An interesting note is that in 1953 when we experienced a freak high tide, the Cliffe Marshes became flooded, many grazing sheep were drowned and the rotting vegetation and carcasses still lying about, a sudden chemical change in the water also killed hundred of eels that were living in the disused clay basins. This greatly concerned the A.P.C.M. and they carried out a number of laboratory tests to find out the cause and to see that no foreign matter was in the clay. Fortunately their concern proved negative.

By the 1960's the A.P.C.M. were now digging inland for supplies of clay: one such site was at Cobham, Kent, where there were large deposits of clay in one of the chalk saucers formed in the chalk hills of the North Downs.

The clay was first transported to the two Medway works of Halling and Snodland by heavy lorries but now the clay is transported by ariel ropeway. A ton of clay is carried in each bucket, the empty buckets returning on the other side. This system makes for a cheap and efficient overland means of transport.

On the Thames, they are now digging London clay at a site at South Ockenden in Essex, but this method of transport is still to pump the clay from a washmill to a jetty and be shipped over to Kent by the clay carrying vessels, but at the time of writing, it was being considered to lay an iron pipe across the Thames on the river bed which would supply one of the largest cement works in Europe which is the new modern works of Bevans at Northfleet.

Some of the barges that the 'Combine' used to run the clay to the Thames works, were now becoming very old and wet, so it was decided to line the barge's holds with a steel tank. Of the final fourteen craft doing the claywork, five had their holds lined with steel, and they were: *Alliance I, Crow, Dorothy, Hevelyn* and *Resurga*. The other nine were *Alice & Ella, Alliance II, Emu, Goshawk, Loualf, Swale, Teal, Tinzanhorn* and *Victoria*. With this number of craft it was possible to get a continual turnround of craft which were at this time running clay to Johnson's, White's and Bevan's works, all in close proximity of each other.

At the same time, however, Brice had the contract to supply clay to Pearsons 'Wouldham Works' at Grays, Thurrock in Essex, which was built to supply cement for the building of the new Tilbury Docks. He employed about nine craft to do this work, it being only a short run from Gray's to the steam grab at Higham Bight.

The craft doing this work were *Busybody, G.C.B., Silica, London Belle, Nelson, Coombdale, Pioneer, Plover* and *Mariana*, all fairly old vessels at this time, but the crews earned very good money getting in three freights a week.

Also at this time, the firm of F.T. Everard had a contract to supply clay to the 'Tunnel' cement works at Grays in Essex.

THE 'COMBINE' CLAY BARGES CIRCA 1930s

Barges with tanks fitted and the amount of tons carried.

Alliance I	105 tons
Crow	130 tons
Dorothy	130 tons
Helvellyn	120 tons
Resurga	150 tons

Barges with-out any tanks fitted and the amount of tons carried.

Alliance II	120 tons
Ally Sloper	115 tons

Alice & Ella	115 tons
Emu	125 tons
Goshawk	115 tons
Loualf	130 tons
Swale	115 tons
Teal	125 tons
Tinzanhorn	150 tons

A.P.C.M. LIGHTERS

LIGHTERS USED IN THE CLAY WORK

Name	*Reg. Ton.*	*Official No.*	*Port of Registry*	*Where built*	*Year*
Bosnia	98	135244	London	Millwall	1913
Inula	98	132661	London	Millwall	1912
Jasminium	98	132674	London	Millwall	1912
King Cup	99	132679	London	Millwall	1912
Lupin	99	132689	London	Millwall	1912
Mimulus	99	132690	London	Millwall	1912
Nasturtium	99	132691	London	Millwall	1912
Whitlavia	99	132735	London	Millwall	1912
Xinga	99	132736	London	Millwall	1912

GOOLE LIGHTERS

One	90	115837	Goole	1,000 casks	1901
Two	90	115836	Goole	of cement	1901
Three	90	115926	Goole	”	1901
Four	90	115838	Goole	”	1901

From the 1900's the A.P.C.M. began widely using lighters for the transportation of materials.

The first were some of the big Goole lighters which were used on the river Medway. These lighters could carry upwards to a 1,000 casks of cement and could be towed two or three at a time by tug.

Later they had a fleet of clay carrying lighters working to the mouth of the river Medway.

These lighters were towed by the firm's own tugs.

CEMENT BARGES

Around the 1900–1910 era, the upper reaches of the Medway were extremely busy with the barge traffic working at this time.

With each tide one could see the coming and going of at least forty barges.

Craft coming up river would be deep loaded with raw material for the cement industry and craft sailing down would be loaded deep with the finished product.

But also in this era there reigned a hidden discontent amongst the barge captains. Although there prevailed a thriving industrious atmosphere in the cement industry, the freight rates were competitive. Even in those early years there was slight inflation and there had been no increase in the barges freight rates for a very long time.

There were however, skipper owners of barges who were satisfied with the way things were as they made a comfortable living.

The situation did not change until the bargemans' strike of 1912 when the general freight rates for barges was increased. On the 1st January 1912, an official list of all freight rates was distributed to all members of The Union of the Amalgamated Society of Watermen, Lightermen and Bargemen.

So, for the coming few years, things seemed to be financially happy and remained industrious right through the 1st Great War to the 1920's. It was about this time that things started declining in the barge world very rapidly. The barges were old and so were the experienced skippers. It seemed the time for change.

By 1910 also, the 'Combine' or A.P.C.M. had really established itself.

The combine now had a huge fleet of barges to repair and maintain. When they went on the 'ways' the barges were painted in the traditional A.P.C.M. colour scheme which was the black hull, grey rails, green transon and the spreet painted with bands of white-red-white-red-white at the sling. In the mainsails were the initials A.P.C.M.

The barges that were owned by the B.P.C.M. however had exactly the same colour scheme on their hulls, but a different set of colour bands on the sling of the spreet, which was red-white-red-white-red and carried the initials B.P.C.M. in their sails. Finally when it became all 'Combine' property, the colours of the bands on the spreet were changed to white-blue-white-blue-white, which was said was due to the fact that blue clay and the white chalk made cement.

The barges at this time were sporting the blue circle in their sails and it was understood to take not less than 1cwt of colour to paint on.

The barges that were incorporated in the 'Red Triangle' had a red triangle in their sails; this was cut out of linen and sewn on. These motifs in the sails gave them instant identification of ownership of up to two miles.

The barges working out from the many wharves on the upper Medway would leave at various stages of the ebb. The craft working out of Burham were usually the first to leave, next would be the barges out of Halling, then Wouldham, all these getting mixed up with the craft coming down from Maidstone, all beating down river on the fast flowing ebb.

Barges being huffled through Rochester bridge 1905. The barge on the left is. loaded with coke, note the hatches stood on edge. The stumpy barge is believed to be carrying clay. Note – each has towing two dingys.

HUFFLERS ON THE RIVER MEDWAY

In the lea of the Temple Marsh would wait the numerous 'Hufflers' or river pilots, sitting in their 'skiffs' patiently waiting for their allotted craft to come down river on the ebb.

Once they spotted their barge, they would row or scull out into the stream to meet her and be in a good position to get alongisde. They usually threw a line to the mate aboard the barge, then pulled in along-side to climb up the lee-board letting their boat go aft on a long painter. This could always be a sign that a barge had got a 'Huffler' on board if the barge was trailing two boats.

Once aboard the 'Huffler' and the barge's mate would go forward to start lowering down the barge's gear which consisted of topmast, main-mast, sails and standing rigging: this all added up to the weight of up to 5½ tons.

The skipper stayed at the wheel to navigate the barge through the main span of Rochester Bridge, which was a hundred and seventy feet wide, and the main granite buttresses, one hundred and forty feet long. Very often this manouvre was made extremely difficult by the terrific pressure of the land water flowing into the river far upstream, particularly in winter or after a heavy storm.

This sudden charge of landwater flowing into the river, added by the natural landfall along its course, caused a six inch fall in the level seawards from Borstal down to the Rochester Bridge Hole.

This water pressure caused numerous eddys in the current, sometimes causing the barge to wallow about as she went through the arch, and it was not an uncommon occurrence for a craft to go diagonally and sometimes catch the rudder or her head on the buttresses as she went through.

Very seldom did the master of a barge let the 'Huffler' take the wheel for they believed that if their barge was doomed to hit a bridge, it was better for the skipper to have done it at the wheel and be responsible, than for the 'Huffler' to have done it at the wheel and the skipper to be responsible. The master of a barge is always responsible for any damage incurred to his craft, be it through accident or by his, or anyone else's negligence.

Once the barges were through, they would moor in Bridge, or Limehouse reach where the gear would be raised back up and the craft 'trimmed' ready for sea.

Many of the skippers would go ashore at Rochester for provisions: they would scull over to the Ship Pier and once their shopping was done would have a bit of a 'booze-up' in the 'Ship' Public House, a favourite meeting place for many of the cement-barge skippers. The more sophisticated barge-masters would use the 'North Foreland' Public House.

Two other popular stops the cement-barge skippers made were at Gillingham Pier where they would congregate in the 'Admiralty Tavern' and the 'Port Victoria' on the Isle-of-Grain, where things were always a penny over the top of the amount they spent. Once all their money had gone, there was no holding them to London, all racing for turn.

At the height of the cement industry, skippers and mates alike made good money, if this was any consolation for always being away from home.

A freight of cement from Burham to Chelsea in 1912 was worth £8.6s.8d. and a freight of coal back could be worth another £5.13s.4d. This money, after the river dues were paid, was divided between master and mate on a three fifths to skipper and two fifths to mate basis.

This was not really a lot of money in view of the fact that both men may have had wives and families to keep, but they were happy times and there always seemed to be an abundance of orders.

Barges that were beating up the River Medway would pick up a

'Huffler' who would be waiting in his skiff at a point in the river known locally as Convict Yard. When they recognised their allotted barge coming up Chatham Reach, they would get out in the main-stream ready to get aboard and once aboard, would start to lower the gear down enough to clear the arch of Rochester Bridge.

When a barge was about a hundred yards from the bridge, the skipper would swing the wheel over to starboard and cant the barge to west-ward then aided by the small bridge sail and the swift current, the barge would be taken up through the bridge. Once the craft was clear the barge mate and 'Huffler' would start the laborious task of winding the tons of gear back up. The 'Huffler' would collect his fee from the barge captain which before 1910 was 1s.6d. for a stumpie, and 2s.0d. for a topsail river barge, however these rates were very competitive: it has been known for a 'Huffler' to take the price of a couple pints of beer.

As the craft got nearer to Rochester Bridge, they would have a tendancy to bunch up at an area below the Temple Marsh and some nasty mishaps occurred when barges fouled Rochester Pier.

This was helped out by The Medway Conservancy who put down several peg-top buoys for a craft to get a few quick turns and bring themselves up on. Photo no page

By the 1920's it was the usual thing for the barges to lower their gear upstream and to be towed down through Rochester Bridge six barges at a time by one of Knight's tugs *Kite, Kennet* or *Kenley*.

The advantages of these towing facilities was that barges could be towed through in both directions at all states of the tide.

Hufflers

The term 'Huffler' was applied to a man who assisted the crew of a barge to lower down the masts and gear enabling it to go under or 'shoot' a particular bridge on a river, then to assist in winding it back up again once they were clear of the bridge hole.

This manouvre called for the utmost skill, and a 'Huffler's' knowledge of all tide and weather conditions on this stretch of the river was invaluable to the masters of barges and their services were much sought after, particularly if the weather was rough and squally.

A 'Huffler' was a very special breed of man, tough and a wily type of character, out in all sorts of weather, at all hours, waiting in his skiff for barges coming along on the tide. He usually spent the waiting time fishing, repairing nets or splicing ropes.

They were usually ex-bargemen themselves and knew every trick in the trade, also they were some of the best barterers on the water.

As the number of sailing craft dwindled, so did the 'Hufflers' who were forced to seek other employment.

With the advent of the steam tugs for towing the need for 'Hufflers' was finished.

The pretty little mud barge *Miranda* in the 1927 Medway barge match.

Part II: Barge Owners

THE EARLY YEARS OF BARGE OWNING

In the very early years of cement making, the mills would only grind around fifty to a hundred tons per week; much of this would be taken away by carts and used locally. Occasionally a barge would pull in alongside the wharf for a freight of cement of between 150–200 casks and take it to London.

As the cement industry expanded they needed a much more efficient system of water transport and so the idea was born for cement firms to build their own craft at their works and they soon built up sizeable fleets of barges, of which they were very proud.

Some of the first barges were part skipper – part company owned and had regular employment with the firm. One of the first names that comes to mind is the form of 'Lee's' at Halling. Barges had been built at Halling as early as 1859, then Samuel Smith built the *Snodland* for Willima Lee – this barge was probably one of the early 'Swimheaders'. Smith built a whole succession of sturdy little craft for the cement industry, some were 'Swimheaded' others were round bowed and they were narrow in build which made them suitable for canal work.

The later additions to the fleet were good sea-going craft and were of composite construction and built at Cowes, Isle of Wight.

The firms fleet were treated like yachts, never overloaded or rough handled. They were brightly painted, had black hulls, a grey transom and rails, red ochre deck dressing, black sails and a blue spreet. At the sling were bands of red-white-yellow-white-red, the name and port of registry was done in goldleaf. The craft that traded with the firm of Eastwood the brickmakers sported 'Horsa' the rampant horse of Kent and 'Lee & Eastwood' in their sails. The firm also had a steam-boat built at Cowes which was named *Lee*. The firm build a fleet of up to 34 vessels, nearly all tiller steered. The next firm which comes to mind is The Burham Brick Cement and Lime Company which had their

Burham works furthest up the river in Lord Oak Reach. This Company was founded by the Peters Brothers and some of the Margetts Family which owned The West Kent Portland Cement Company, at about the same time. The B.B.C.L. Co. barges were a little on the small side, very good for river work and were stumpy rigged. They did however have several topsail river barges and these ran most of the clay to the works.

The firm had dredged a channel up to their works which short-circuited a sharp bend in the river at this point. This gave them a greater depth of water at their river wharves than they would otherwise have had.

The firm also had a cement works at Murston near Sittingbourne but this was sold away to Mr George Smeed around the 1870's. The B.B.C.L.Co. built up a sizeable fleet of some 34 sturdy barges, very handy little craft, which were either built at Murston, Aylesford or Burham by their own shipwrights. Their barges were distinguishable by their black hulls, green transoms, grey rails and carried 'B & Co' in their sails.

They were named after the months of the year, days of the week, four seasons and the quarter days.

Next comes the firm of The Peters Brothers whose fleet comprised of some thirty barges some of which were sizeable craft but nearly all were stumpies.

A few of the fleet namely *The Ninety Nine, Alumina* and *Silica* were built especially for claywork to the factory.

Their craft were distinguishable by their black hulls, black rails, brown quarter boards and transoms and had a 'P' in their sails.

Another of the big names in cement manufacturing about this time was the firm of Trechmann Weekes & Co., who established a sizeable cement works on the banks of the river Medway in the early 1900's. This was in Whornes Place Reach, Halling. The firm gradually built up a small fleet of barges which they acquired from various sources but all adding their contribution to the ever demanding cement industries barge traffic.

Yet another of the big cement manufacturers at the turn of the century was Hilton-Anderson & Brooks. Philip Hilton had established works at Faversham and Upnor, Herbert William Anderson at Halling and Edmund William Brooks at Grays, Essex. These had vast experience of cement making and together built a large works at Halling Reach on the Medway.

The firm built up a fleet of thirty four barges most of which were bought from other firms at barge auctions.

Another quite prominent cement works on the River Medway in the 1900s was owned by the West Kent Portland Cement Company which was founded by the Margetts family and partners in the B.B.C. & L. Co. Apart from their manufacturing of a good quality cement, they also produced a fine gault brick by machine.

The firm's small fleet of barges were splendidly maintained, they were painted all black including rails and transom and they carried a six pointed white star in their topsails. The *Britannia, Hibernia* and *Caledonia* were rather bigger than the normal run of cement barges. The remainder of the fleet were named after members of the Margetts family. The fleet was completed by the part skipper owned *Outsider*.

This is an outline of the Fleet Lists of barges that were owned by cement manufacturers on the River Medway between the years circa 1850–1930.

THE PHOENIX PORTLAND CEMENT COMPANY

Name	*Reg. Ton.*	*Official No.*	*Port of Registry*	*Where built*	*Year*
Florence	34	4574	Rochester	Lambeth	1855
Hawk	43	78509	Rochester	Frindsbury	1877
Phoenix	64	90970	Rochester	Frindsbury	1884
Cerf	58	90988	Rochester	Frindsbury	1886

WILLIAM TINGEY – FRINDSBURY KENT

Name	*Reg. Ton.*	*Official No.*	*Port of Registry*	*Where built*	*Year*
Robert Bladen	33	9444	Rochester	Frindsbury	1847
Richard	30	28534	Rochester	B. Stortford	1849
Charles	30	12590	Rochester	Greenham	1854
Eliza	41	47941	Rochester	Frindsbury	1863
Gwendoline	44	84444	Rochester	Rochester	1882
Alice & Ella	52	86612	Ipswich	Wandsworth	1882

JOHN BAZLEY WHITE & CO
THE BRIDGE CEMENT CO.

Name	*Reg. Ton.*	*Official No.*	*Port of Registry*	*Where built*	*Year*
Dispatch	35	11223	Rochester	Maidstone	1838

FORMBY CEMENT CO. WHITEWALL, FRINDSBURY, KENT.
(Jas. and Chas. Formby)

Name	*Reg. Ton.*	*Official No.*	*Port of Registry*	*Where built*	*Year*
Sarah	39	7995	Rochester	Frindsbury	1845
Pink	43	20063	Rochester	Frindsbury	1847
Queen	43	7994	Rochester	Frindsbury	1848
Halling	38	7969	Rochester	Halling	1849
Protection	38	20068	Rochester	Halling	1850
Neptune	40	20069	Rochester	Frindsbury	1853
Whitewall	37	52950	Rochester	Frindsbury	1865
Frindsbury	42	52963	Rochester	Halling	1865

Ex-Lees *Medina* entering Portsmouth Harbour.

Ex-Lees *Medina* hulked at Fawley, Southampton Water 1968

Vauxhall	40	58440	Rochester	Frindsbury	1867
Eclipse	39	58481	Rochester	Frindsbury	1869
Kew	40	81847	Rochester	Kew	1879
Daisy	43	81855	Rochester	Rochester	1879
Margaret Louise	45	87204	Rochester	Frindsbury	1882
Ella Vicars	43	87208	Rochester	Frindsbury	1883
Ida Gibson	40	90977	Rochester	Rochester	1885

Situated on the north bank of Chatham Reach on the River Medway.

WILLIAM LEE AND SONS

Name	*Reg. Ton.*	*Official No.*	*Port of Registry*	*Where built*	*Year*
Conservative	39	27614	Rochester	Halling	1859
British Queen	35	44591	Rochester	Halling	1862
Snodland	38	44618	Rochester	Halling	1862
Superb	38	47947	Rochester	Halling	1863
Rover	38	52938	Rochester	Halling	1865
Defiance	39	52962	Rochester	Halling	1865
Henry Tinnoth	50	55141	Rochester	Halling	1866
Renown	38	55183	Rochester	Halling	1866
Invicta	40	58455	Rochester	Halling	1867
Telegraph	39	58470	Rochester	Aylesford	1868
Ariel	37	67081	Rochester	Rochester	1874
Graphic	40	74807	Rochester	Rochester	1876
Punch	40	76586	Rochester	Rochester	1876
Judy	42	76601	Rochester	Rochester	1877
Urgent	41	76629	Rochester	Rochester	1877
Holborough	41	81893	Rochester	Rochester	1880
Shark	44	84383	Rochester	Cowes	1881
Dolphin	43	84386	Rochester	Cowes	1881
Porpoise	44	84397	Rochester	Cowes	1881
Medina	42	84412	Rochester	Cowes	1881
Dart	43	84418	Rochester	Cowes	1881
Thames	44	84428	Rochester	Cowes	1881
Medway	45	84435	Rochester	Cowes	1882
Tyne	42	87211	Rochester	Cowes	1883
Maidstone	42	–	Rochester	Rochester	1891
Blackfriars	42	98800	Rochester	Rochester	1891
Lee (steel yacht)	67	76625	Rochester	Cowes	1877

Situated at Halling Reach on the River Medway
(Those built at Cowes were of composite construction.)

The ex-B.B.C & L Co. *September* a complete write-off after a collision in the Thames. Circa 1929.

A complete cross section of the *September* after she was cut down in the River Thames circa 1929.

THE BURHAM BRICK CEMENT & LIME CO.
(Burham & Wouldham)

Name	*Reg. Ton.*	*Official No.*	*Port of Registry*	*Where built*	*Year*
Edward	41	18270	Rochester	Strood	1857
James	42	45519	Rochester	Frindsbury	1863
Esther & Jess	40	49849	Rochester	Chatham	1865
Ann	40	55151	Rochester	Frindsbury	1866
John	40	55173	Rochester	Frindsbury	1866
Murston	35	58436	Rochester	Sittingbourne	1867
Varnes	41	58450	Rochester	Frindsbury	1867
Venus	37	58452	Rochester	Sittingbourne	1867
Surrey	37	58457	Rochester	Murston	1868
Ann	36	58463	Rochester	Murston	1868
William	41	58467	Rochester	Frindsbury	1868
Meteor	36	58490	Rochester	Murston	1869
William Porter	62	58529	Rochester	Battersea	1871
Gun	44	45511	Rochester	Frindsbury	1875
Autumn	43	84378	Rochester	Aylesford	1880
Sussex	45	84434	Rochester	Rochester	1882
Monday	47	84443	Rochester	Burham	1882
Tuesday	42	87222	Rochester	Aylesford	1883
Wednesday	41	90973	Rochester	Burham	1885
Thursday	44	94557	Rochester	Aylesford	1888
Friday	46	108520	Rochester	Halling	1865
Saturday	46	110964	Rochester	Rochester	1899
Sunday	46	110969	Rochester	Rochester	1899
January	36	58477	Rochester	Sittingbourne	1868
February	35	58483	Rochester	Aylesford	1869
March	38	58485	Rochester	Rochester	1869
April	41	5846	Rochester	Rochester	1869
May	41	58491	London	Greenwich	1869
June	35	58497	Rochester	Murston	1869
July	37	58502	Rochester	Boston	1869
August (1)	66	67058	Rochester	Aylesford	1873
August (2)	44	90987	Rochester	Aylesford	1886
September	46	67089	Rochester	Aylesford	1874
October	40	67118	Rochester	Aylesford	1876
November	45	76593	Rochester	Aylesford	1876
December	43	78520	Rochester	Burham	1878

Sited at Burham Old River.

Peter's Works Wouldham, looking up river.

Peter's Works Wouldham, looking down river.

WOULDHAM & BURHAM CEMENT WORKS

Henry Peters – Cement & Lime Manufacturers

Name	*Reg. Ton.*	*Official No.*	*Port of Registry*	*Where built*	*Year*
Providence	33	25473	Rochester	Rochester	1823
Joseph Livesey	45	–	Rochester	Strood	1846
Elizabeth	43	4923	Rochester	Strood	1846
Edwin	46	7997	Rochester	Rochester	1854
Ellen	44	9416	Rochester	Rochester	1855
John	38	47951	Rochester	Frindsbury	1864
William	39	47956	Rochester	Frindsbury	1864
Harry	39	52955	Rochester	Rochester	1865
Peters	32	55149	Rochester	Rochester	1866
Providence	38	74819	Rochester	Rochester	1876
Alice	42	76590	Rochester	Rochester	1877
Marie	43	76591	Rochester	Rochester	1877
Enterprize	41	81860	Rochester	Cowes	1879
Beagle	42	81894	Rochester	Rochester	1880
Foxhound	42	81895	Rochester	Rochester	1880
Mersey	42	84406	Rochester	Rochester	1881
Joseph	41	84407	Rochester	Rochester	1881
Scarborough	41	87205	Rochester	Rochester	1882
Ninety	43	999&7730	Rochester	Rochester	1890
Wouldham Hall	44	109927	Rochester	Rochester	1898
Wouldham Court	44	109928	Rochester	Rochester	1899
Overcomer	44	109929	Rochester	Frindsbury	1899
Monkwood	46	109930	Rochester	Frindsbury	1899
Silica	54	110961	Rochester	Sittingbourne	1899
Alumina	60	110962	Rochester	Sittingbourne	1899
Edwin	40	118203	Rochester	Sittingbourne	1903

Sited on the south side of the River Medway at Peters Reach.

HILTON, ANDERSON AND BROOKS & CO.

(Proprietors:– Philip Hilton, Upnor – Herbert William Anderson, Halling and Edmund William Brooks, Grays.)

James & Emma	31	4333	London	Northfleet	1846
Henry & Jane	32	23314	London	Vauxhall	1847
Alice	44	21885	Faversham	Faversham	1859
John	49	44596	Rochester	Strood	1862
Julia	40	49833	Rochester	Rochester	1864
Anne Maria	44	50303	Rochester	Faversham	1864

Thames	36	50306	Faversham	Faversham	1864
Medway	39	50308	Rochester	Faversham	1865
Richard	35	65771	Rochester	Chiswick	1867
Annie Kendall	41	58453	Rochester	Upnor	1867
John Little	42	58459	Rochester	Upnor	1868
Hope	44	28849	Rochester	Milton	1877
Clyde	41	87221	Rochester	Rochester	1883
Helvelyn	49	87228	Rochester	Sittingbourne	1884
Cader Idris	50	90969	Rochester	Sittingbourne	1884
Forth	42	90975	Rochester	Rochester	1885
Daisy	43	52905	Faversham	Faversham	1886
Plimlimon	48	91902	London	Strood	1886
Tinzanhorn	50	96696	London	Blackwall	1890
Mendip	82	98137	London	Bow Creek	1890
Mersey	44	89866	Faversham	Faversham	1891
Mahatma	52	99008	London	Bow Creek	1891
Ural	80	101907	London	Canning Town	1892
Swale	48	104932	Rochester	Faversham	1894
Humber	46	108512	Rochester	Sittingbourne	1897
Dee	46	109914	Rochester	Sittingbourne	1898
Tay	46	109915	Rochester	Sittingbourne	1898
Shannon	57	109920	Rochester	Sittingbourne	1898
Dart	39	110957	Rochester	Faversham	1899
Emily Stewart (Schooner)	113	36245	Rochester	Holland	1855
Cotswold (Converted Lighter)	78	98137	London	Blackwall	1887
Cheviot (Lighter)	84	95442	London	Rochester	1888
Chilton (Lighter)	76	95445	London	Blackwall	1888
Grampian (Converted Lighter)	78	95494	London	Blackwall	1888

Sited on the north bank of Medway at Halling Reach.

BOOTH & CO., BORSTAL, KENT.
Arthur William Booth – Nashenden

Name	*Reg. Ton.*	*Official No.*	*Port of Registry*	*Where built*	*Year*
Gertrude	41	67075	Rochester	Rochester	1874
Maud	38	79872	Rochester	Milton	1875
Nashenden	44	108517	Rochester	Milton	1898
Lancelot	41	58504	Rochester	Rochester	1870

Sited at Borstal reach on the south side of the River Medway.

TRECHMANN WEEKES CO., HALLING, KENT

Name	*Reg. Ton.*	*Official No.*	*Port of Registry*	*Where built*	*Year*
Expedition	30	25454	Rochester	Maidstone	1792
Murcury	29	18269	Rochester	Newbury	1824
Busy Body	37	14479	Rochester	Aylesford	1837
John Tinnoth	43	4574	Rochester	Frindsbury	1851
William & Sarah	41	18076	Rochester	Frindsbury	1853
George	45	20043	Rochester	Frindsbury	1855
Ambrose	40	20571	Rochester	Frindsbury	1858
Two Brothers	35	27619	Rochester	Halling	1860
East Kent	38	44095	Rochester	Sittingbourne	1862
Henry Tinnoth	50	55141	Rochester	Halling	1866
Bella	35	74808	Rochester	Frindsbury	1876
Edward & William	40	78530	Rochester	Frindsbury	1878
Stratford	42	81899	Rochester	Frindsbury	1880
Estelle	44	110956	Rochester	Rye	1899

Sited on the north side of the River Medway at Whornes Place Reach.

WEST KENT PORTLAND CEMENT CO. BURHAM

Name	*Reg. Ton.*	*Official No.*	*Port of Registry*	*Where built*	*Year*
William & Martha	43	25487	Rochester	Strood	1852
Ethel Margetts	48	76617	Rochester	Strood	1877
Stanley Margetts	44	79878	Rochester	Frindsbury	1878
Arthur Margetts	43	90979	Rochester	Aylesford	1885
Harold Margetts	45	97717	Rochester	Frindsbury	1890
Cecil Margetts	46	97718	Rochester	Frindsbury	1890
Outsider	49	104316	Rochester	Rochester	1895
Hibernia	45	110967	Rochester	Strood	1899
Caledonia	46	110959	Rochester	Rochester	1899
Britannia	46	113682	Rochester	Rochester	1900

Sited on the south side of the River Medway at Lord Oak Reach.

THE FALCON CEMENT WORKS, UPCHURCH, KENT
(Frederick C. Barron)

Name	*Reg. Ton.*	*Official No.*	*Port of Registry*	*Where built*	*Year*
Snodland	38	44618	Rochester	Halling	1862
Iolanthe	43	89549	London	Greenwich	1884
Liqtar	43	89580	London	Millwall	1884
Nonpariel	50	110965	Rochester	Rochester	1891
Noclaf	45	98807	Rochester	Rochester	1891
Norrab	45	99926	Rochester	Rochester	1893
Veronique	56	120510	London	E. Greenwich	1905

Sited on the east bank of Otterham Creek.

This is a brief description of Fleet Lists of barges owned by cement manufacturers on the River Thames between the years 1850–1930.

ROBINS & CO. NORTHFLEET
Manager – Robert Andrew Gibbons

Name	*Reg. Ton.*	*Official No.*	*Port of Registry*	*Where Built*	*Year*
Mary	38	9962	London	Lambeth	1837
W.H.F.	35	56803	Rochester	Millwall	1867
Alliance	50	67095	Rochester	Rochester	1874
Countess	37	67110	Rochester	Rochester	1875
CLipper	42	67104	Rochester	Rochester	1875
Kate	47	76583	Rochester	Rochester	1876
Adelaide	48	74812	Rochester	Rochester	1876
W. Paxton	52	84377	Rochester	Gravesend	1881
J.M.W.	44	84417	London	Blackwall	1881
R.A. Gibbons	55	85161	London	Northfleet	1882
Louisa	44	87121	London	Northfleet	1883
Bride	42	98678	London	Milton	1885

PADDLE TUGS OWNED BY ROBINS & CO

Jessie May	58	73669	London	Poplar	1876
Pearl	51	76938	London	Cubbit Town	1877

GIBBS & CO., GRAYS, ESSEX

Three Brothers	35	50095	London	Chiswick	1863
Lion	44	56888	London	Millwall	1867
Lioness	39	60930	London	Millwall	1869
Alliance	43	85083	London	Brentford	1881

Francis & Co. of Cliffe, Kent

The firm of Francis & Co were another of the early pioneers of cement making.

The manager and part owner of the Cliffe cement works was Gerald B. Francis who was the son of Charles Francis who, together with John White, formed a small cement firm as early as 1825, known then as Francis & White. This firm was dissolved by 1827.

John White took his knowledge of cement making to the Isle of Wight where he founded the Medina Cement Company, while Charles Francis formed a company and built the Nine Elms Cement Works at Cliffe.

This works at Cliffe expanded to some considerable size and it thrived right up until the formation of the 'Combine'.

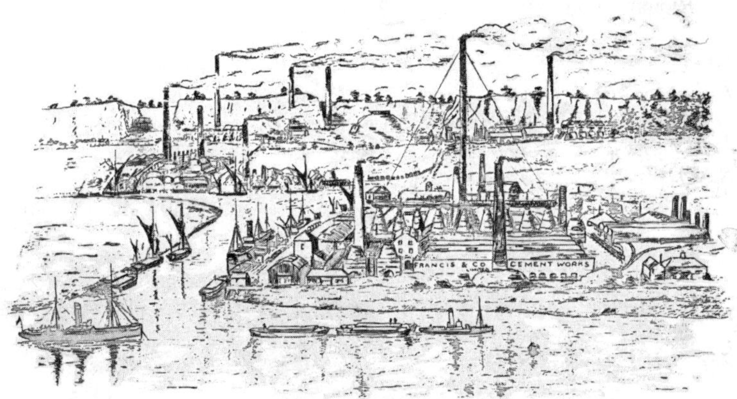

GENERAL VIEW OF THE NINE ELMS CEMENT WORKS, CLIFFE, KENT.

Large quantities of Ferro cement was supplied from this works for the building of several lighthouses.

The firm had a London office and wharves at Nine Elms. Their motif depicted nine elms which was displayed on the sails of their barges.

FRANCIS & CO., CLIFFE, KENT'
Gerald B. Francis

Eddystone	50	94579	Rochester	Rochester	1889
Minicoy	50	96690	London	Rochester	1889
Lizard	50	99023	London	Rochester	1891
Needles	49	99042	London	Rochester	1892

Richard Tolhurst of Northfleet

Richard Tolhurst was another of the very early cement manufacturers of Northfleet who helped pioneer the wholesale production of cement before the turn of the century.

He built a fairly large works at Northfleet right on the river's edge. The basic raw material chalk was close at hand. It was dug from the 'Bluffs' behind the works.

The firm owned six sailing barges on record, and these craft took away the finished product from the 'Red Lion Wharf' and, all important, brought in the much needed supplies of clay.

Richard Tolhurst formed the Imperial Portland Cement Company. Their 'Flag' brand cement was of a very high standard.

The firm was absorbed into the 'Combine' in the 1900s. When the old works was demolished it became the site for the then new Imperial Paper Mills.

THE IMPERIAL PORTLAND CEMENT COMPANY
Richard Tolhurst, Northfleet Works, Red Lion Wharf, Gravesend

Gundolph	44	67097	Rochester	Frindsbury	1875
Indian Empress	51	79868	Rochester	Kew	1878
John Pain	61	84422	Rochester	Brentford	1881
Rathmona	45	106529	Rochester	Milton	1886
Silver Wedding	50	106526	Rochester	Rochester	1896
Dorothy	49	110116	Rochester	Limehouse	1899

A view of the John Bazley White's Works Gillingham Kent. Note – the barge unloading at the mud-berths extreme left of picture.

John Bazley White

The firm of John Bazley White was one of the earliest of Portland Cement manufacturers on both the River Thames and Medway. Having first established a cement mills at Swanscombe in the 1850s, they went on to build several more on the River Medway, one at Gillingham and three at Frindsbury.

The firm were always keen to take advantage of the very latest technical and technological advances that took place within the industry, and so were able to keep pace with the increasing demand for more cement. Bazley White's were among the first cement manufacturers to have their own bargebuilding yard and to build many of their own barges.

They built a yard at Swanscombe at an area known locally as the Black Duck where they had a slipway and a set of barge blocks.

The very first barge to leave the ways was a barge named the *Black Duck* which was in 1882. After which they built several other topsail river and coasting barges named after birds, *The Swan*, 1882, *Cygnet* 1883, *Gannet*, 1884 and *Dabchick*, 1885.

Before this time J.B. White had several other craft built at Greenhithe by Alfred Keep. All these craft were very large, and had a good carrying capacity and could carry well over 150 tons to sea.

The largest craft of all that was owned by the firm was the *J.B.W.* which was built of steel at Millwall in 1907.

THE TUNNEL PORTLAND CEMENT WORKS CO.

Henry & Mary	39	55180	Rochester	Chatham	1866

JOHN BAZLEY-WHITE – SWANSCOMBE & NORTHFLEET

Caroline	43	9402	Rochester	Rochester	1854
Edward & Charles	43	28533	London	Queenborough	1860
Emily	51	29381	London	Lambeth	1860
Flower-of-Kent	44	49839	Rochester	Frindsbury	1864
Lucy	43	52678	Rochester	Lambeth	1865
Fanny	41	81825	Rochester	Rochester	1870
Little Emly	54	70609	London	Greenhithe	1874
George	42	87052	London	Greenhithe	1874
Harry	49	87053	London	Greenhithe	1874
Laura	45	67093	Rochester	Millwall	1874
Sarah	38	23425	Rochester	Frindsbury	1874
A.K.	53	77051	London	Greenhithe	1876
Loo	52	87051	London	Greenhithe	1876
New Ada	48	76604	Rochester	Milton	1877
New World	–	76624	Rochester	Milton	1877
Victoria	45	108239	London	Greenhithe	1879
George & Jane	35	81879	Rochester	Millwall	1880
Sapphire	51	84226	London	Landport	1880
Hilda	72	87054	London	Henley	1881
Black Duck	58	86982	London	Swanscombe	1882
The Swan	59	87050	London	Swanscombe	1882
Cygnet	57	87118	London	Swanscombe	1883
Bittern	60	89494	London	Swanscombe	1883
Moorhen	63	·89555	London	Greenhithe	1884
Gannet	61	89617	London	Swanscombe	1884
Dabchick	62	89670	London	Swanscombe	1885
Teal	53	91890	London	Swanscombe	1885
Garland	44	90998	Rochester	Rochester	1887
Irex	44	94577	Rochester	Rochester	1889
Teal	53	91890	London	Swanscombe	1895

Regent	35	105805	London	Swanscombe	1896
Nan	47	105890	London	Swanscombe	1896
Sea Gull	54	108350	London	Greenwich	1896
Plover	62	110026	London	Swanscombe	1898
J.B.W.	72	123813	London	Millwall	1907

THE EAST GREENWICH PORTLAND CEMENT WORKS

Three Brothers	35	50095	London	Chiswick	1863

Bevans of Northfleet

Another of the very early cement makers at Northfleet was one Thomas Bevan.

Thomas Bevan established his first cement works in this area in the 1860s. This early works consisted of a few bottle or beehive kilns. Bevans have always kept pace with modern cement making techniques. In 1906 they had some of the first rotary kilns installed at their Northfleet works, and from that time they have kept pace with progress.. At this present time the A.P.C.M. have built one of the most modern and automated cement works in Europe on the same site that Bevan used over a century ago.

Thomas Bevan was one of the cement manufacturers who also built and maintained his own barges. He established a barge yard right alongside the works. The first barges on record to come off the ways were *Portland* and *The Hive* in 1862. He then went on to build a whole succession of really good sea-worthy craft that were mostly named after adjectives or birds.

It was the opinion of some that Bevans barges were always being very hard driven. It has been known for a barge to be on the blocks for a scrape and tar-up, this being done in drizzling rain or when they would be loading loose lime down a shoot on a windy day at Fletchers 'hit or miss' wharf a few hundred yards away, and then without a minute to lose the barge would be taken straight round to be loaded with a freight of bagged cement in some little tucked away fore and aft dock by the works.

The last two on record to be built by Bevan were the big old *Quail* and the *Ibis* in 1892. Both these craft could carry a freight of nearly two hundred tons.

When the bargeyard closed it was only used for repairs. Bevans craft were always distinguishable by their orange-tinted sails; this colour was made by simply adding a drop of yellow-ochre to the sail dressing.

Several of the firm's bigger barges ran to the mudholes on the Medway, namely they were the *Kestrel, Goshawk, Merlin* and the *Crow.*

In later years Bevans had four old 'Nipcat' barges that were towed about to do this work.

BARGES OWNED BY THOMAS BEVAN OF NORTHFLEET

Active	40	63554	London	Northfleet	1869
Alert	40	63664	London	Northfleet	1870
Swift	41	65622	London	Northfleet	1871
Brisk	41	65738	London	Northfleet	1872
Success	40	68494	London	Northfleet	1873
Despatch	42	73607	London	Rochester	1875
Progress	41	73604	London	Northfleet	1876
Prompt	42	73743	London	Rochester	1876
Spry	44	77094	London	Northfleet	1878
Falcon	59	87088	London	Northfleet	1883
Kestrel	63	89503	London	Northfleet	1883
Goshawk	66	89535	London	Northfleet	1884
Merlin	62	89635	London	Northfleet	1884
Raven	64	94387	London	Northfleet	1888
Crow	60	101911	London	Northfleet	1891
Jay	60	101910	London	Northfleet	1892
Quail	79	102819	London	Northfleet	1893
Ibis	75	101996	London	Northfleet	1893

JOHN MESSER KNIGHT OF NORTHFLEET
(Incorporated with Thomas Bevan, became
known as Knight Bevan & Sturge)

Alert	40	63664	London	Northfleet	1870
Belvedere	37	43998	London	Bankside	1862
Breakwater	37	44852	London	Southwark	1862
Conservative	39	27614	Rochester	Halling	1859

Part III: Cement Works

This is a brief list of some of the many cement works that were established on the banks of the River Thames and River Medway before and after the 1900's.

MEDWAY WORKS – NORTH KENT AREA (1910)

Falcon Cement Works, Otterham Creek (1889–1914 – mud from saltings nearby); Sharps Green C.W.; Lower Upnor P.C.W.; Gillingham P.C.W. Gillingham Creek; Whitewall C.W. Upnor.

ROCHESTER – ALL ON N.E. BANK OF LIMEHOUSE REACH

Phoenix C.W.; Globe; Bridge; Crown; Quarry; Beehive; Beaver; Wickham C.W.; Borstal Manor C.W.; Borstal C.W.; Medway C.W. Cuxton; Trechmann Weekes; Halling Lime & C.W.; Clinkham Lime Works (later Rugby P.C.); Wouldham Lime & C.W.; Manor Works, Halling; Halling Lime & C.W.; Wouldham Hall Lime & C.W.; West Kent P.C.W.; Burham Brick, Lime & C.W.; West Kent C.W.
NB Holborough Works, Martin Earles and British Standard not yet in existence.

THAMES WORKS

Nine Elms; Swanscombe; Britannia; Portland; Tower; Portland (on Ebbsfleet Creek); Northfleet; Crown; London; Imperial Red Lion; Albion; Shield; Greenhithe; Johnson's; Dartford (North of town on east bank of Darenth).

SHARPS GREEN CEMENT WORKS

About the same time as the Queenborough Cement Works was thriving, another small cement works was operating in conjunction with Alfred Castle.

This cement works was built of second-hand machinery at the head of The Sharps Green Peninsula, and was owned by Francis Joseph Carey of Greenhithe together with Jos. Wilders of Stone. The works consisted of a bank of small chamber kilns, a tall brick chimney, a bagging shed, and a grinding mill powered by a gas engine.

The works employed the minimum of men on a shift system. The supplies of chalk for this works were supplied by Alfred Castle from the pit at lower Gillingham.

Although there were adequate beds of clay near to the works, the barges *Edith* and *Chancellor* brought the supplies of London clay from Stoke where thre was always a gang of 'muddies' to dig and load.

This works was absorbed into the 'Combine' and closed around 1910.

THE SITTINGBOURNE WORKS

Over the past hundred years there have been a number of cement works in Sittingbourne, particularly in the area of Crown Quay.

One of the first to manufacture cement there, was a man named John Huggens of Northfleet.

Around the 1860's he built several beehive kilns on the banks of

Muddies 1972 at the Churchfields mudhole in Milton Creek. L to R photographer Alan Cordell, Clifford Willmott, and Author Frank Willmott.

Milton Creek, together with a clinker grinding mill, This was known as The Sittingbourne Cement Works.

Another works that opened up in this area a few years later, was Bulivants Cement Works, under the auspices of Messrs Clever & Mist.

In the 1880's another works was operating in this area, this was built and worked by Edward Rosher who had a successful works operating at Northfleet. The works at Crown Quay consisted of six or more beehive kilns.

Just before the turn of the century, Charles Burley built a cement works on the banks of Milton Creek, just below Crown Quay. This was a fairly modern works for that time, built more like a factory. The slurry was fired in a batch of horizontal 'chamber' kilns. There was also a grinding mill, a bagging shed and several handy wharves to serve the works.

This works was known as The Dolphin Cement Works, and their 'Dolphin Brand' cement becamed renowned on the London markets. The clay supplies to this works were dug from an area further down the creek on the opposite bank at Churchfields, where Burley had a clayhole.

The clay was dug by hand and loaded into several of Burleys older barges which had had their gear taken out and were used as lighters. Once these craft had been loaded, they were poked up to the works with long setting booms to be unloaded at the clay berth at the works.

This Dolphin Cement Works thrived up until 1928 when it was closed.

SMEED-DEANS

The firm of The Burham Brick, Cement & Lime Company established a small cement works at Murston in the east bank of Milton Creek as early as 1860–70 together with George Smeed who built several barges for this company.

It was after this date that the B.B.C. & L. Co. moved out from the area leving George Smeed who had now gone into partnership with Dean to form the well known firm of Smeed Dean.

The cement works was rebuilt and modernized.

Chalk supplies to this works were pumped through a six inch diameter pipe from Highsted where there were large deposits of pure chalk.

As far as any records go, the clay for this Murston works was dug by hand from the many clay beds along the Swale, mainly from a point above the 'Hutch'.

Two of the early barges that did this work, were the *Monitor* and the *Perseverance*.

These barges were loaded by hand by a gang of 'Muddies' who worked on a weekly wage basis rather than on piecework. The foreman

The Smeed Dean Works, looking up Milton Creek.

The Smeed Dean Works, looking down Milton Creek.

'Muddie' was Tom Hammond together with brothers Peter, Jim, Buff, George and Bill. Also in the gang were Bill and Jack Shilling, and Jack and Tom Reed.

These men would go down to Elmley Ferry cross onto Sheppey, walk down to the mudhole, load the barges waiting there and then return by the same route. This arrangement lasted for years.

In later years this works operated under the 'Red Triangle' Group. At this time there were now some four barges doing the mudwork, they were *Easthall* Captain Ernie Spice, *Sam* Captain Frank Farrington, *Winnie* Captain Harry Freeborne, and *Livingstone* Captain Harry Port.

The barge *Russell* was also doing this work at sometime. When the firm was absorbed into the 'Combine' in 1930, they were loaded by an A.P.C.M. Grafton Steam Crane on a pontoon which was moored in the Swale. The barges doing this work were now towed about by the A.P.C.M. tugs *Kismet, Leopard* or *Panther*.

When these craft all became too old for claywork and had to be scrapped, the 'Combine' began digging for clay inland about half a mile away from an area known as 'Bloombanks'. Here they dug the clay by grab and loaded it away in side-tipping trucks – these were towed by loco, ten or twelve at a time to the works.

Later the firm dug clay from the head of Milton Creek, where they began digging and forming clay basins.

The Murston Works was closed by the A.P.C.M. in 1970 when the modern Northfleet Works was working at full production, and eventually was demolished in 1974.

SMEED DEAN BARGES THAT DID THE CLAYWORK

Name	*Reg. Ton.*	*Official No.*	*Port of Registry*	*Where Built*	*Year*
Easthall	40	99930	Faversham	Murston	1893
Sam	39	104326	Rochester	Murston	1895
Winnie	45	81896	Rochester	Murston	1880
Livingstone	48	81883	Rochester	Murston	1869
Russell	44	79895	Rochester	Murston	1879
Perseverance	39	60892	London	Northfleet	1868
Monitor	41	44100	Faversham	Faversham	1862

CASTLE'S QUEENBOROUGH CEMENT WORKS

On the Isle-of-Sheppey were to be found two more small cement mills.

One of the works was sited at the head of Queenborough Creek which made it accessible to both road and river. This was started in the early 1860's by a man named Alfred Castle. By 1890 it was known as the Queenborough Cement Works and consisted of a cooper's shop, small bagging shed together with four beehive kilns producing sometimes 400 tons a week, and a grinding mill.

Although supplies of clay were dug from the saltings nearby, the chalk supplies were bought in by barge from as far away as Gillingham.

It was dug from a pit at Lower Gillingham and then taken by rail in side-tipping trucks, to a high wharf on the west side of the Sharps

Green Peninsula and was tipped down a large wooden shoot into Castle's own barges and taken to Queenborough. The Queenborough Cement Works thrived up until 1912 – the works were demolished during the 1914–18 war to provide hardcore for the war department roads being built on the Island. Alfred Castle owned a number of barges which ran the cement away to London and beyond.

An interesting gact is that Castles ran thousands of tons of flint and chalk to Sheppey to help build up the railway embankment for the extension of the London, Chatham and Dover Railway to Queenborough in 1860.

QUEENBOROUGH CEMENT COMPANY BARGES
Alfred & James Castle

Name	*Reg. Ton.*	*Official No.*	*Port of Registry*	*Where built*	*Year*
Elizabeth	39	45528	Rochester	Sittingbourne	1863
Emma	41	55135	Rochester	Sittingbourne	1866
Peacock	41	55162	Rochester	Rochester	1866
Alice	46	67111	Rochester	Millwall	1875
Trent	42	104328	Rochester	Frindsbury	1878
T.F.C.	46	104323	Rochester	Rochester	1895
Thelma	49	113701	Rochester	Sittingbourne	1901

THE TURKEY CEMENT WORKS

The other cement works on the Isle-of-Sheppy was at Elmley, opposite the mouth of Milton Creek on the Swale.

This works was built in the 1860's and worked by Messrs William Levett and Co and included several other big names in cement manufacturing at that time.

At this cement works they produced a 'Roman' cement. This 'Roman' cement was made by burning nodules of natural argillaceous chalk found in the London clay of this locality, in this case they were the saltings of the Sheppey side of the river Swale.

Levett had a little swim headed barge that was regularly employed running the clay to this works, she was named *Admiral Blake*. The little village of Elmley grew up around the cement works and when it closed just before the turn of the century, the inhabitants moved onto the mainland.

When the Turkey Cement Works closed in 1900, William Levett together with other interested partners had considerable cement making interest on the Frindsbury Peninsular.

THE FALCON CEMENT WORKS – F.C. BARRON

This fairly large cement works of its time was sited on the east bank of Otterham Creek around the 1900's. It consisted of several banks of chamber kilns, a grinding mill, a cooper's shop, and several large wharves that served the works.

The clay that was used at the works was dug from some saltings on the opposite bank of the creek.

It was brought over to the works in a lighter named the *Harriet*. This lighter was poked about with a long pole. It was taken over to the salting at high water to a mud berth and once the tide had gone out, was then loaded by hand by a gang of eight muddies, four each side, and would load a hundred tons during low water. As soon as there was enough water for it to float, it was poked over to the works and unloaded.

When this lighter became too old for the job and the saltings were dug away, the firm employed a barge to run the clay from the 'Bogs' in East Rainham Creek.

The barge that ran the clay was named the *Snodland*, she also carried a freight of 100 tons.

The chalk for the works was pumped through a six inch iron pipe from a pit about two miles and a half away.

The basic fuel coke for the works was brought down from Beckton or East Greenwich Gas Works in two other barges owned by the firm *Norrab* and *Noclaf* which are backslang names of Barron and Falcon.

This works was absorbed into the 'Combine' around 1910, and was closed a couple of years later.

THE BRITISH STANDARD CEMENT WORKS – RAINHAM KENT

E.J. & W. Goldsmith of Grays

The name of Goldsmith ranked high amongst the top names of barge owners on the Thames and Medway.

This firm of barge owners and carriers was started before the 1880's and they went on to build up a huge fleet of sailing barges, ranging from small stumpies 'swimmies' topsail–river up to the large steel coasters of 300 tons burden that were built in Holland at the turn of the century. The firm boasted of owning the largest fleet of sailing barges on the two rivers which comprised of some 130 craft at one time, excluding the numerous lighters which the firm owned.

An arial photo of the British Standard Cement Works at Rainham, Kent. The works were built in 1913 for E.J. & W. Goldsmith of Grays in Essex. Note – in the top right hand corner the clay breakwater of the Bogs clay hole.

Around 1909 the firm decided to have built a large cement factory at Rainham in Kent.

This modern works was opened in 1913 and consisted of two large rotary kilns that were automatically fed with dried cement slurry and coal dust.

There was a large grinding mill at the works and ample storage space and loading facilities which were all under cover.

The works had large docking facilities for the many sailing barges that ran to the works.

The works was officially known as The British Standard Cement Works and came within the 'Red Triangle' group.

The supplies of chalk for the works were dug from a pit about a half a mile away and was transported to the works by locomotive and side tipping trucks on a network of rails that ran around and in the works.

The clay was dug from the 'bogs' near the works, but in later years was brought over from the Stoke Marshes by barge. At the end the barges *Gloria* and *Haughty Belle* were doing the claywork, and it was on that fateful day in 1929 that the *Gloria* capsized with the loss of both hands, in a fierce squall. The works thrived up until the 1930's when it was finally absorbed into the 'Combine' and declared uneconomical and subsequently closed.

A MUD BARGE TRAGEDY

One of the most serious of tragedies that befell the mud barges was the sinking of the Goldsmith ironpot *Gloria* when she capsized in a fierce squall in Kethole Reach in the River Medway, resulting in the loss of both her crew by drowning.

The precise date of the accident was the twenty-ninth of December 1929. The *Gloria* was running supplies of clay into the British Standard Cement Works at Rainham, Kent and the weather at this time had been very rough for several days but the craft working to the clayholes still managed to keep working as always. Just prior to the sinking, the *Sphere* and the *Ella & Norman* were being loaded at Kingsnorth in one of the mudholes, while the *Gloria* and the *Haughty Belle* were being loaded in another. As she drew just that much less water, the *Sphere* came out of the mudhole just before the *Ella & Norman* and as it was nearing dusk and the light was beginning to fade, together with thunder and lightening with hail and a fierce gale blowing, it was the decision of the skipper of the *Sphere* to run for the shelter of Damhead Creek and lower the gear down.

As they were coming down river they passed the *Gloria* and *Haughty Belle* coming out of Stoke Creek. The crew of the *Sphere* laid in Damhead Creek all of Friday and Saturday and even so had dragged her

Mark (Curly) Ransley skipper of the Goldsmith Iron-pot *Gloria* which capsized in a squall in December 1929

anchor a few feet.

As the weather calmed down and the barges began to get under weigh again, it was learned that the ***Haughty Belle*** had run up into Otterham Creek with just her topsail and foresail set and that they had passed the ***Gloria*** at the Bishops Spit Buoy so that was the last anybody had seen her afloat.

Unloading clay at Anchor Works

General view of Anchor Works from River Thames

The mystery of how the *Gloria* capsized remained unsolved; her skipper Mark (Curly) Ransley was a very experienced man on this part of the water and he had worked as a muddie on all of the saltings in the area at some time or another.

Captain Ransley's body was recovered a few days later, but the mate's body was never recovered.

The *Gloria* was raised up and beached at the Rainham Cement Works where she was broken up for scrap.

BROOKS SHOOBRIDGE OF GRAYS ESSEX.

This firm of cement manufacturers was established in 1870 and their fairly modern works for that time, was sited on the north bank of the River Thames at Grays in Essex, close to abundant supplies of the purest chalk.

The proprietors of the firm were some of the early pioneers of technical methods of cement manufacture.

At their 'Anchor' works at Grays, they perfected some mechanical methods of intermixing and washing of the basic raw materials i.e. chalk and clay, also of the burning of dried slurry and grinding of the resulting clinker.

The firm was amongst the first to take periodical samples of these materials to keep a watch on the consistency of their quality.

Because of these methods, their 'Anchor Brand' of Portland Cement became unsurpassed in its purity, and maintained a long reputation for uniform in character, good colour and of high tensile and rapid hardening qualities.

This enabled it to be held in very high esteem by constructors of dry docks, reservoirs, culverts, sewers, wharves and embankments, both at home and abroad.

Thousands of tons of this cement was taken up river by barge and lighter to be shipped aboard the foreign-going vessels berthed at the London docks.

Part IV: Barges and Freights

BARGE BUILDERS

One of the little known barge builders on the River Medway in the 1870s was William Highams who established a bargeyard at Rochester.

Highams yard was at a point on the river known locally as Blue Boar Soft. Any traces of the yard have long since gone and hardly any records exist of the work carried out.

Highams built most of his barges on spec. many of which were bought by young skippers on a sort of hire purchase system, so much deposit and so much a freight was paid back per month or week.

The skipper owners of these barges would seek regular employment from the many little cement firms that were operating at this time on the banks of the Medway.

When several of the small mills were bought out or closed or absorbed in the 'Combine' many of these barges went over to the combine as well.

Many of the Highams barges were absorbed into the A.P.C.M. fleet in this way. Most of the craft were stumpy rigged.

Here is a list of barges that were built at Rochester by William Highams and were absorbed into the A.P.C.M. Fleet.

WILLIAM HIGHAMS & SONS, BARGE BUILDERS,
BLUE BOAR HARD, ROCHESTER

Name	*Reg. Ton.*	*Official No.*	*Port of Registry*	*Where built*	*Year*
Providence	38	74819	Rochester	Rochester	1876
Alice	75	76590	Rochester	Rochester	1877
Foxhound	42	81895	Rochester	Rochester	1880

Joseph	41	84407	Rochester	Rochester	1881
Scarborough	41	87205	Rochester	Rochester	1882
Florence Myall	58	90968	Rochester	Rochester	1884
Ally Sloper	60	97724	Rochester	Rochester	1890
Herbert & Harold	42	98793	Rochester	Rochester	1890
Answers	45	104312	Rochester	Rochester	1894
Silver Wedding	50	106526	Rochester	Rochester	1896
Victoira	44	108338	London	Rochester	1898
Duplicate	50	110020	London	Rochester	1898

BARGE ROADS

The 'Combine' also needed large mooring buoys for the craft to moor up to, these were called 'Barge Roads' or 'Tiers'. These needed to be in a sheltered spot near the works out of the main fairway and where there was a good depth of water.

On the Medway, these buoys were at Rochester, Frindsbury, Chatham and Upnor.

On the Thames there were many more; working up the river they were at Northfleet, Grays, Woolwich, Blackwall, Greenwich, Millwall, St. Pauls, Lambeth, Nine Elms, Battersea and Chiswick.

BARGE YARDS

Together with good mooring facilities the 'Combine' ran several efficient 'Barge Yards', which had a good slip or 'ways' together with a grid or 'blocks' where a craft could receive a good overhaul or have her bottom scraped and a coat of tar.

These barge yards were once barge building yards and on the Thames were at Grays, Northfleet, Swanscombe and Greenhithe.

On the Medway they were at Burham, Halling and Upnor.

During my research, I cam across a list of some of the 'Combine' barge yard employees together with a list of some of the old skippers. Whether these employees were shipwright riggers, blacksmiths or sailmakers, they left real epitaphs to their high degree of skill in their own particular job and it's true to say that some of the finest examples of workmanship in blacksmithery and shipwrighting remain. However, neither they or the bargemen left an epitaph to their fine seamanship or knowledge of survival and self sufficiency.

With the 'Combine' having around three hundred barges, they must have had nearly that number of skippers and mates, not forgetting the sailmakers, blacksmiths and shipwrights to maintain a large fleet.

Captain Vic Wadhams, served the A.P.C.M. for fifty years mostly on barges.

During my research, I came upon many names that were associated with the barging side of the A.P.C.M. and which I have listed below. Many of the names are repetitive where fathers, sons, uncles and brothers were on the water leaving one in doubt about christian names, but they all semed to have a nickname by which they were known.

When the 'Combine' started selling off the barges they had in trade, most of the skippers went with their barge. Other employees went into agriculture or the local factories and mills.

Now that all the barges have gone, our loss is some of the finest examples of blacksmithery, shipwrighting and sailmaking ever seen. Not only was each craft a one-off job but a creative design for the men building her based on their previous experience to see whether they could make further improvements to shape, carrying capacity, manoeuvrability and speed.

Each man carried out his job with pride, from the laying of her keel to the fitting of the sails.

BARGEYARD

G.E. Adsley
F. Barritt
G. Gulivan
M.G. Scotchman
W. Tapsell
C. Taylor

SAILMAKERS

G. Gozmarks
J. Packman
A. Smith

A.P.C.M. BARGE CAPTAINS

F. Butler
R. Ball
J. Ballard
Benson
Bennetts
A. Chandler
A. Chapman
W. Coleman
C. Coleman
J. Dawkins
J.S. Dandy
R.W. Fuller
S. Flint
Fuller
E.W. Gray
G.H. Hanslow
Hickford
C. Hutson
E. Lutchford
W. London
J. Dewsberry

W. Inge .
C. Inge
C. Money
D. Morgan
G. Marshall
H. Nicholas
H. Pound
C. Redding
J. Sutton
W. Scott
E. Simmonds
J.F. Sayers
I. Wicker
C. Woolmer
W. Webster
R. Wooding
C.H. Wooding
R. Wilkinson
W. Wadhams
V. Wadhams

FREIGHTS

The cement and lime trade made more use of the sailing barge than any other single industry. Over the years millions of tons of these commodities were handled by these fine little craft.

Cement and particularly lime was no cargo for a leaky barge and so the barges were by necessity of the soundest construction and kept in first class condition.

Lime

Lime could be a very dangerous cargo as in early days it was shipped as 'quick-lime' and usually loaded straight from the kilns. Any water seeping through into it would start a slaking process generating sufficient heat to char wood or even start a fire, and it was part of a lime barge's equipment to carry a two inch diameter auger so that in the event of water coming in or a fire starting on board, a few holes could be hastily drilled in the cabin or fo'castle floor to scuttle the barge and save her from fire. A barge carrying a freight of cement was sometimes spared from an ordeal like this, as the cement was usually shipped in casks.

A freight was usually around 500 casks or not quite 90 tons.

Coke

Coke was the main fuel used in cement and lime making so most craft that worked a freight to the London river or docks would proceed on unloading to one of the big gasworks at Beckton or Greenwich for a return freight of coke. When a barge took on a freight of coke, it was a considerably lighter commodity and to get a good sizeable freight on, it was usual to load all holds in the normal way, and then batten down the forehatch, but the hatch covers of the main hold were arranged round the coamings to take on more of this much lighter cargo, after which her hatch cloths were lashed down overall. A few feet of the mainsail were taken up to clear the extra height of the freight.

CLAY PRICES FROM 1912–1936

1912

From any clayhole in the Medway to works below Rochester Bridge, including Otterham ..... 6d per ton

Above Rochester Bridge ..... 8d per ton

Minimum freight 80 tons.

From any clayhole in the Medway to works below Dartford

Dartford	.....	10d per ton
Dartford & Greenwich	.....	1s. 0d per ton
From Burham to Grays	.....	1s. 4d per ton
Higham Bight to Dartford Creek	.....	7d per ton
From Dagenham to below Dartford Creek	...	8d per ton

1920 Revised

From any clayhole in the Medway to Rainham (Kent) and

Otterham	.....	9d per ton
Below Rochester Bridge	.....	10d per ton
Above Rochester Bridge	.....	1s. 0d per ton

Minimum freight 80 tons.

From any clayhole in the Medway to works below Dartford

Creek	.....	1s. 0d per ton
Higham Bight to below Dartford Creek	...	10d per ton
Higham Bight to Greenwich	.....	1s. 0d per ton

Minimum freight 80 tons.

These prices were revised again in 1936.

1936

Any clayhole in the Medway to

Below Rochester Bridge	.....	1s. 5½ per ton
Above Rochester Bridge	.....	1s. 9d per ton

Minimum freight 80 tons.
Demurrage £1.15s per day

Any clayhole in the Medway to works below the entrance to Dartford

Creek	.....	1s. 6d per ton
Dartford to Greenwich	.....	1s. 9d per ton
From Burham to Grays	.....	2s. 3d per ton
From Higham Bight to below Dartford Creek	...	1s. 3d per ton
Higham Bight to Greenwich	.....	1s. 6d per ton

Demurrage £1.10s per day.

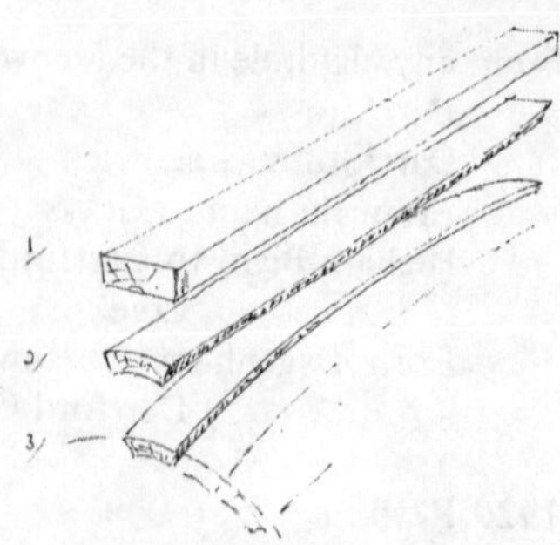

COOPERING

In the early days of cement making, the loose cement was usually put into barrels or casks for transportation to various destinations.

Watertight barrels or casks kept the cement dry if it was to be stored for any length of time, or if the casks were to be transported by barge.

Coopering was a considerably large allied industry to cement manufacturing, and it was not uncommon for a cement works to have their own coopers shop situated in the factory complex.

The average spritsail sailing barge would carry around 500 casks of cement, so it was important to produce large quantities of empty casks in relation to the output of cement that was being ground.

It was not unusual in a large cement works to employ as many as 50-60 men and boys in the coopers shop.

Some casks made for exportation of cement to the Indian Government, had as many as six wooden hoops and three iron bands on the same cask.

The casks for the Crown Agents however, had all iron hoops and the staves were tongued and grooved: each and every cask was made to a very high standard. An empty cement cask weighed approximately 24lbs and was filled up with 404lbs of cement before the head was finally tapped home and the last three hoops tightened on.

On the head of each cask was burned the port of destination, i.e. Port Sweetenham, Port Ceylon, Port Calcutta and Port Rangoon were but just a few.

Barges would take freights of cement to the London Docks to be loaded aboard any of the British India P. & O. or Well boats for their long journey overseas.

There was, however, a marked difference between barrels and casks:

Barrels were made and used to transport the export cement where it had to stand a long journey and perhaps to be handled up to three or four times en-route. These barrels were also constructed to a very high standard.

Casks were usually of a lighter construction and would not have so many hoops around the bilge. Cement for the home market was transported in these casks.

In the early days of cask making the rough staves were imported from Norway and Sweden.

They were brought to the London Docks by three or four masted Schooners.

The cement barges would often pull alongside one of these vessels for a freight of staves, these would be brought down to the works where there was a wharf alongside the coopers shop, or a drying shed for the staves. The rough staves were twenty-eight inches long and in various widths of two, two and a half, three, three and a half and four inches. It was of the utmost importance that these staves were kept free from moisture content before the casks were assembled. Gross shrinkage of the staves after the casks were filled, resulted in the loss of cement, particularly in a warm climate and where the casks were rolled about en-route to the construction sites.

The 'Anchor' works on the river Thames was one of the first works in England to artificially dry the staves for cask making. They were also one of the first to introduce the tongueing and grooving of the staves and headings of barrels by machine and double riveting of the steel hoops. These improvements meant that the cask and its contents reached its destination in excellent condition and soon other works began copying their example for an improved type of barrel.

Nearly all the tasks of the coopering trade and throughout the cement industry were done at piecework rates. Before the first world war, women were employed to unload the staves which were carried to the store and stacked in their various widths, five at a time. The women who did this work were paid about 2s.6d. a day.

The barrel or cask making was usually done in a clockwork system and was completed in three different stages. Firstly there was what was known as a 'Bellman' whose job it was to make up the frame of the barrel and give it its right diameter and shape etc., then it was passed on to a man known as the 'Hooper' whose job it was to fit all the required number of staves together in the frame and to ensure all the joints were not too tight or slack. The barrel was then passed on to the 'Cooper' who assembled it together, tapping on the required number of iron and wooden hoops.

With this system going, it was possible for each cooper to make an average of thirty casks or barrels a day.

The ends of the staves above the throat were bevelled so that when the barrel was filled with cement, the head could be tapped home. As this was done, it would spirng open the staves at the end so that the head would clip into the circular rebate and once the head was in this rebate, the last three hoops would be tapped on.

Before the 1900s, the staves for cask and barrel making were made in the coopers shop at the cement mills, or by a contractor in the village.

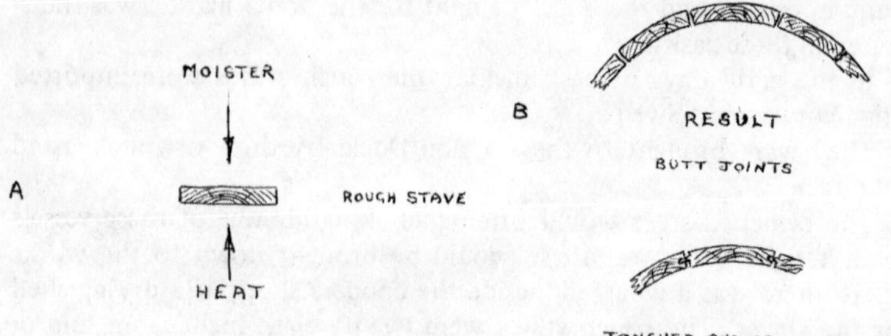

The staves were fashioned from pieces of oak board whcih measured approximately four inches wide by one inch thick, and to get part of the required shape the rough boards were put through a sort of wet and drying process. To do this they were laid on a slated iron grill with the smallest diameter annual rings facing down.

A considerable amount of heat was applied underneath the grill from the coke ovens of the cement kilns, and on top of the boards water was gently applied – these two processes together caused the boards to curl which was just the right arc for a barrel stave.

By 1912, jute sacks began to be widely used on the home market for the bagging up of cement. These sacks were made from a tightly woven jute material so that the cement did not sift through and were never completely filled so when used the cement could be tightly stowed in the barge's hold. Each sack of cement weighed approximately 1cwt. and was known as a 'Pudding bag'.

There was however a larger type of jute sack which held considerably more cement, approximately 1¾cwt, and went eleven bags to a ton.

This method of transporting cement went on up until about 1929-30, when the three and four ply paper sacks began to be used. This proved a more economical method of bagging as paper sacks were not returnable as in the case of the jute sacks, casks and barrels – these were all returned, repaired and re-used.

CEMENT KILNS

Beehive Kilns

Many of the early cement 'Kilns' or 'Cowls' were very simple in construction.

In the early 1850s the most popular type of kilns were the 'Bottle' or 'Beehive' kilns, so called because of their shape. These kilns were sometimes built in a line or in banks of fours, sixes or eights and were built in fire brick to enable them to withstand the terrific heat that was built up during the firing.

Old bottle kiln.

Some of these kilns or cowls were twenty feet in diameter and almost the same height.

At the face of each cowl was the eye or entrance to a fire-bricked floor on which the layers of raw materials were laid. First to be laid was a layer of coal or coke, then a layer of cement slurry, then more coke until the kiln was topped up. The loading of the kiln was done from the top through square ports on the sloping side of the cowl.

A series of temporary walkways were rigged up for the loading. In an arched duct under the brick floor were placed suitable combustible materials, e.g. faggots soaked in oil or paraffin, to start the initial firing. Once the cowl was set alight the whole lot was allowed to burn through.

Once it had burnt through and had cooled down enough, the remaining cement clinker was broken out, loaded into carts and taken to the mills for grinding into fine powdered cement.

Intermittant horizontal kiln.

Chamber Kilns

By the 1880s many large cement companies had built chamber kilns at their works.

This fairly simple but advanced type of kiln produced considerably more cement clinker than the earlier types. The chambers of these kilns were constructed in fire bricks which made them easy to repair when large cracks appeared due to expansion during firing, or a roof collapse during the breakout.

The kilns were built in banks of threes, i.e. nine, eighteen or twenty-seven. Each bank of three chambers was a hundred foot long, twenty feet wide and eighteen feet high.

At one end of each chamber were two openings, one at floor level and one about nine feet up the face. Acess to the upper one was from a steel plate that ran across the face of the chambers, wide enough for two barrows to pass. This was used during the breaking out of the clinker. Once the chamber had been emptied both entrances or front eyes were bricked up. Reloading was done from the upper access ports, or eyes, located at regular intervals along the chambers length.

Running under the kilns below ground level was an arched tunnel four feet high, this was the actual furnace. To fire the kiln, bundles of faggots soaked in paraffin were placed in the furnace tunnel and ignited.

Between the furnace and chamber was a ventilated fire brick floor. On this floor within the chamber was laid another layer of faggots and then a six inch layer of coke.

Behind the chimney was built a large wet slurry tank under which were rows of ducts which carried the hot air exhaust from the chamber to dry the wet cement slurry before passing up the chimney. This hot air exhaust was used to accelerate the drying process of the slurry, which was difficult to do during the winter months.

Rotary Kilns

This now very modern method of firing the raw ingredients, was first experimented with as early as 1896 when a rotary type kiln was built at Grays in Essex. This was not an instant success as a lot of technological factors had still to be worked out.

The experimentation of this method was completed in America and in 1900 the first successful rotary kiln was brought over from the States and erected at the new Martin Earles 'Wickham' factory on the River Medway.

The rotary kiln proved to be an instant success and the firm gave orders for the installation and building of a battery of sixteen more of these kilns to be installed at the works. It is true to say that this method of producing raw cement clinker is still used to this day.

LOADING GANG

The loading up of a chamber kiln was done by a gang of six men on a piece-work basis.

This gang consisted of two 'Breakers', two 'Humpers', and two 'Loaders'. All of these men were hard manual workers, each could do the other's job when required.

To load a chamber, a six inch layer of coke was put down on the floor with a layer of faggots, then alternate six inch layers of dried cement slurry and coke were placed until it built up to within eighteen to twenty inches from the top.

All this loading was done from the top of the chamber, the raw materials were tipped through rows of 'eyes' or ports. Alongside of these ports were placed temporary wooden walkways for the loaders.

The gang of six men would split up; the two 'Breakers' would be in the coke yard to load the baskets for the 'Humpers', each basket of coke contained about three bushels of coke, these were lifted onto the 'Humpers' backs by the 'Loaders'; he would be 'landing', that is, to take down the basket of coke and tip it into the eye.

In the chamber would be the other 'Loader' who would be raking and spreading the materials to an even thickness. Each layer of coke took nine to ten rounds per man which was approximately eighteen to twenty bushels of coke.

The dried cement slurry (chalk and clay) was loaded into barrows from the slurry pans at the same level as the top of the chamber and were sited at the rear of the chambers.

It was wheeled across the top of the bank of chambers on more timbers that ran right across the top to any of the eyes or ports where it was to be tipped. This slurry had been dried by the exhaust heat from the previous firing, and when the pans were emptied, more liquid slurry was run in, the excess water drained off, ready to be dried by the next firing.

Part V: The Clay Work

Most of the clay that is found in the Thames and Medway valleys is the alluvial mud claimed from the banks of the estuaries, from what are known as Saltings.

But some of the clay deposits are part of the London clay beds which are also found inland, and are up to thirty feet thick in places.

This clay is a rich blue-grey colour and contains very little foreign matter, as is sometimes found in the clay dug from the saltings, which contain traces of callow or marsh peat.

This type of foreign matter is considered detrimental to the manufacturing of cement.

In some areas however traces of lime were present.

The use of clay in cement making acted as a flux which helped to fuse the molecules together during the firing process, after which the remaining clinker is ground to a fine powder.

As the ratio for the use of clay in cement making was around five to two, this would be on a larger scale of two hundred and fifty tons of chalk would need an admix of approximately a hundred tons of clay to make it into cement, this amount of clay would be a barge's freight.

So one can get a good idea of how important it is for the cement factories to get their supplies of clay.

Some of the best supplies of good quality clay came from the marshes and saltings of the estuaries of the River Thames and Medway where it has been dug by hand, and by machine for well over a hundred and fifty years. Vast areas were excavated from Motney Hill, Milford Hope, Twiney, Funton, Bishops, Burntwick, and the Hoo Flats, not forgetting the large areas on the Stoke Saltings that were dug away, where at the turn of the century well over a hundred men and boys were employed to dig the clay by hand, and, where fifty to a hundred barges waited their turn to be loaded.

MUDDIES

The term of 'Muddie' was applied to a person who worked at digging clay or mud from the saltings in the River Medway and the lower parts of the River Thames.

This clay was loaded straight into a barge's holds. When the barge had got a full freight of around a hundred tons it was taken to one of the many cement works and unloaded.

In the early days this work was all done by hand, and this loading of the clay was extremely hard work and to acquire the knack for quickness only came from years at working at the job.

An eye witness reckoned that it was a fascinating sight to see a gang of 'Muddies' digging and loading a barge, all in a squence and all keeping to a perfect rhythm, some digging left handed and some digging right handed. The whole gang moved just like clockwork.

Muddies loading *Mahatma* at Stoke Saltings. Note – The lime cloths around the standard rigging and mast case.

It was reckoned that a good muddie if he was quick, could have one spit landing, have one in the air, and be cutting another.

A favourite bet between the muddies was to see who could cut and throw a spit of clay weighing twenty eight pounds up and over the top of a barge. The wager was usually a gallon of beer. The distance it had to be thrown was somewhere in the region of twenty five feet.

One very fast and experienced muddie would always work and load the forward end of the barge before the main gang came to load the barge up. This muddie was known as the 'Hermit'. His job would be to load about twenty to thirty tons of clay in the barge's forehold so that if the tide was making the stern of the barge would lift first and so keep the main weight off the rudder.

A gang of muddies would always work out in any sort of rough weather to finish loading a barge, then they would retire to the warmth and refuge of an old converted lighter or some other type of vessel on which they would sometimes live out on the saltings.

The Muddies were a very hard and tough breed of men which came about by continual hard manual work out in all weathers and seasons.

This type of work soon made them old before their time. They spent their working days amid the creeping insiduous damp of the saltings on which they worked and lived. It soon gave them chronic rheumatism and arthritis.

Some Muddies suffered from sprained wrists and strained backs. Some had large callouses on their hands that sometimes joined several fingers together. This was from the continual using of the spade or 'flytool'.

Another affliction common to the muddie was called 'scaffleman' which was the tearing of the skin on parts of the back, which was the result of lifting, turning and throwing simultaneously. These cracks or tears in the skin were made worse by the sweat from the body and the continual rubbing of the course flannelette vests that they all wore.

All of these complaints resulted in a few days off work. But these men were never shirkers of hard work.

Barges moored up in a clay hole.

THE MUDDIE TRADITION

Like many other jobs around the 1900 era the muddie was very much a father and son tradition.

The muddies earned very good money compared to that of a farm labourer, and they saw to it that it remained a close knit circle. In fact at one time it was said that 'if yer wornt a muddie in Stoke yerd get yer eye cut out wi a quid (Sovereign).

Muddies made very good money when things were going right. Something like four pence a ton was paid for hand loading in the 1900s. The average barge would take about a 100 tons of clay making their earning rate about thirty-five shillings a barge, to be divided amongst eight to ten men. Sometimes they were lucky enough to be able to load two barges every low tide. There were nearly always fifty to a hundred barges waiting on the Stoke marshes to be loaded.

All of the Muddies dressed alike. They wore navy flannelette shirts, corduroy trousers, and a cloth cap. Their footwear consisted of a special type of leather boot which had a long stiffening up the back to give the calf of the leg a bit of protection in case their foot happened to slip off the spade or 'fly tool' which they used.

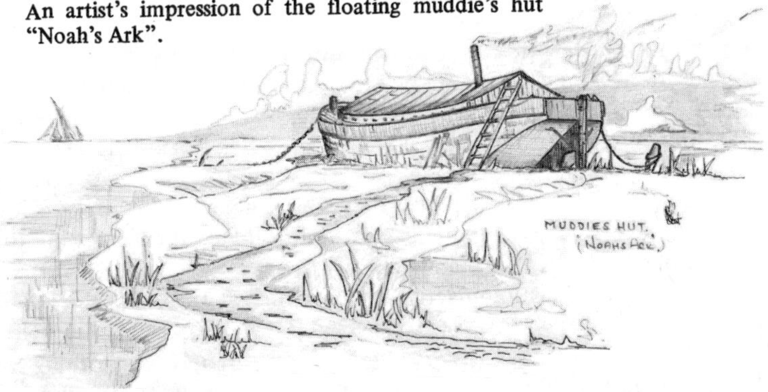

An artist's impression of the floating muddie's hut "Noah's Ark".

With this type of dress they were able to dig furiously, being free from any encumbrant clothing and they would work on in any sort of inclement weather to get a barge loaded and away before they retired to their 'Noah's ark' type of vessel to a warm stove and a meal of bread, cheese, mushrooms and plenty of wild duck's eggs.

Most of the muddies that worked on the Stoke marshes lived in the small hamlets and villages of that area. Large families of muddies lived at Hoo, Middle Stoke and Lower Stoke and it was very seldom that they would let anybody else in to the job.

On oft occasions when the Hoo muddies were going to work on a Monday morning, about twenty to thirty men strong all carrying their bundles of food and personal possessions, one of the gangers

Middle Stoke muddies at Damhead Creek circa 1900

would throw up a brick into the air and if it came down facing frog-upwards they would all hang their bundles on the iron spiked railing fence in the Stoke Road, then they would all go into the public house called the 'Victory' which was in the Stoke Road at Hoo. They would stop in this pub all day drinking, gambling and making merry, nobody daring to go near them, their womenfolk taking cautious looks down the road to see if their bundles were still there, then pass on the message in a hushed voice, "They ain't gorn yet".

The Hoo muddies domineered the 'Victory' at Hoo, and the Stoke muddies domineered the 'Nags Head' at Lower Stoke, and the Hoo brickworkers domineered the 'Bridge House' at Hoo. It was nothing for the Hoo muddies after one of these drinking sessions to go over and fight the Stoke muddies, and when the going got rough would be helped out by the Hoo brickworkers. Other times the Stoke muddies would come over and fight the Hoo brickworkers who would be rescued by the Hoo muddies.

Whichever way it all started they would all finish up drinking together the very best of pals.

I suppose that this was their way of relieving the tensions of their workaday lives, a time when they could relax and rejuvenate their bodies after a good run of work.

When the men returned to work the next day they would work like trojans digging and loading the mud, and feeling much better for their time off.

The Ransley boys, known as the "ungry eight" they would always load two barges each tide. Loading *Eric* away to Wickham's on the bogs at Rainham Kent. Photo circa 1900. Master of *Eric* Captain Ted Phillips.

There were other times when a barge was only half loaded and the tide was making and would cover the clay area, the muddies would would walk up to the pub and have a few pints of old and mild beer until the tide turned, then they would go back and finish loading the particular barge in time for her to just get away.

The money for the men's wages was sometimes brought out to the ganger of the muddies by a man on horseback. It was usually kept safe until it could be divided up to the men. This was nearly always in the pub where they always had a debt to pay for tobacco and beer had on the slate.

A 'Ganger' muddie made extremely good money at that time of the day and when they became old they bought pubs in the area.

Old Sid Mortly bought the 'White Horse' at Upper Stoke and it is still in the family. Old Jack Rapley had the 'Rose & Crown' at Allhallows, and Joe Rayner had the 'Nags Head' at Lower Stoke.

Some of the very familiar surnames among the muddies still survive in the area today, surnames like:– Beeching, Bell, Cheeseman, Collins, Craddock, Driver, Filmer, Leslie, Mudge, Morgan, Mortley and Rayner.

A 'MUDDIE' GANGER

One of the more renowned muddies was old Mr Jim Simmons, who had worked no less than sixty years in the claywork, all of that time with S.J. Brice. Old Jim was a very experienced man at this sort of work and rose up to be Brice's senior ganger and his opinions on things were greatly respected. He used to go down on the Saltings to select areas which he thought would make clayholes. He did this by finding a spot and digging a small hole Then he would reach down with a spoon and then he would actually taste the clay to test its gritty content. Only the best would do!

Jim had seen the good times and the bad times in the clay work and having brought up four strapping sons things were not always easy.

As a point of interest Jim reckoned that he was the nearest man on the Stoke Saltings to the battleship H.M.S. 'Bulwark' when she was sabotaged and blown up in 1914. The terrific explosion threw large fragments of steel and timber in all directions for about a two mile radius.

The firm of Solomon John Brice were very big people in the clay contracting, and almost all of their fleet of up to fifty barges worked in the clay traffic at some time.

They also employed well over a hundred men digging the clay by hand and supplied thousands of tons of clay to the cement works on the River Medway.

By the 1920s many of the cement works had been streamlined and modernised to meet the ever increasing demand for more cement. As they in turn also needed yet larger quantities of clay as the admix to the chalk, so the supplying of this raw material also needed streamlining.

Several well known firms were given large contracts to supply this clay to the works. The firm of S.J. Brice was one of them.

Brice purchased four Grafton steam cranes with grab buckets to hasten the clay digging. These cranes were bolted on to pontoons shaped like a dumb lighter.

One was taken round to Johnson's mudhole on the Lower Stoke marshes, and another was taken round to the Kingsnorth mudhole on the Upper Stoke marshes, and a third was taken round in the Thames and moored in Higham Bight.

The fourth crane was held in reserve at the firm's Point Yard at Rochester. Where once the firm employed over a hundred men to dig the clay by hand, with these four cranes they only needed eight. The rest were made redundant. Most of Brice's barges that were still working in the clay traffic were loaded at the Johnson's mudhole and these took the clay to the cement works on the Medway.

At this time also the 'Combine' still had fifteen barges in the clay traffic and these were now being loaded at the Kingsnorth mudhole and were taking the clay to the cement works on the Thames.

These barges could be loaded by one of these cranes in one and a half to two hours depenidng on the size of the barge. The two largest were the *Resurga* and the *Tinzanhorn* which both loaded a hundred and fifty tons each. The smallest only loaded a hundred and eight tons.

CRANE DRIVERS

Two of Mr Jim Simmons's sons drove Grafton steam cranes for S.J. Brice. His youngest son Joseph, drove the crane at Johnson's mudhole, while his oldest son Solomon, drove the crane at the Kingsnorth mudhole. Solly was one of th first men to drive one of Grafton type steam cranes in 1919. He spent over thirty years doing this job.

Of course the main advantage of a steam crane was that they could load several barges a day.

The clayhole that was being dug away by the crane was up to thirty feet deep and had a small cutting out to the river, anything from two to six barges would go in to mudhole at high water, then as the tide ebbed the barges would remain afloat. Sometimes the first two barges in would get loaded and would get out on the same tide, the others being systematically loaded would have to wait for the next tide, this in turn would give the first two away a chance to get unloaded before the others got there. Among the crane drivers doing this work, Solly Simmons held the record for loading the most barges in a tide. The number he loaded was eleven barges.

Another two records that he held was to dig an estimated four thousand tons of clay in a week paid at a piecework rate of a penny a ton, and to load sixteen barges in succession for £8-4s-0d or a penny a ton. None of these records were ever broken.

Solly Simmons spent a good deal of his time down on the marshes, at all seasons of the year and in all types of weather conditions, and, at all times of the day and night.

There have been a number of times when he has cycled down to the mudhole in the early hours of a summer's morning he could see the barges' spreets sticking up out of the low lying marsh mist.

Sometimes one of the bargemen would shovel some coal into the boiler's fire so that when he got there the boiler would have a full head of steam and he could start loading a barge.

Many of the barges carried on deck a large fresh water tank for the crane to replenish her tank to the boiler.

After clay excavating had finished the mudholes were used as dumping grounds for river mud dredged from the Medway. In 1940–41 many cargoes of bomb debris, rubble and hardcore were taken from London to Kingsnorth mudhole and unloaded by 'Grafton' steam crane.

Four of Brice's barges were employed doing this work. They were the *R.S. Jackson, Glenburn, Plover* and *Alumina.*

THE COMBINE 'MUDDIES'

Some of the 'Muddies' who dug clay by hand for the 'Combine' lived at Frog Farm near Twinney. This gang of 'Muddies' lived in a row of cottages which overlooked the Milfordhope Marshes; they were Frog Farm Cottages and all had mud floors.

The 'Muddies' lived close to the claybeds of Milfordhope Marsh, so were out loading the barges at all hours.

Barges that were owned by the firm of Tuff & Miskin used to run clay from these beds to cement works on the Thames before the 1920's. By 1930 the clay digging by hand had ceased and the digging by 'Grafton' Steam Crane with a grab was the accepted thing. The combine had a crane digging clay from the beds in Twinney Creek, driven by Alf Carlton until his retirement.

When the claybeds began to get a bit low in this area, the A.P.C.M. moved their crane over to Colemouth Creek to dig clay from the 'beds' there. By this time the driver of the crane was Wally Bennett who used to live in a tarred wooden cottage near the old Grain Bridge.

The A.P.C.M. dug clay from this area in a very big way. As many as four at a time of the 'Combine's' clay carrying lighters were brought round from the Thames to be loaded.

The clay digging from this area ceased around 1934 when the A.P.C.M. began digging the clay basins on Cliffe Marshes.

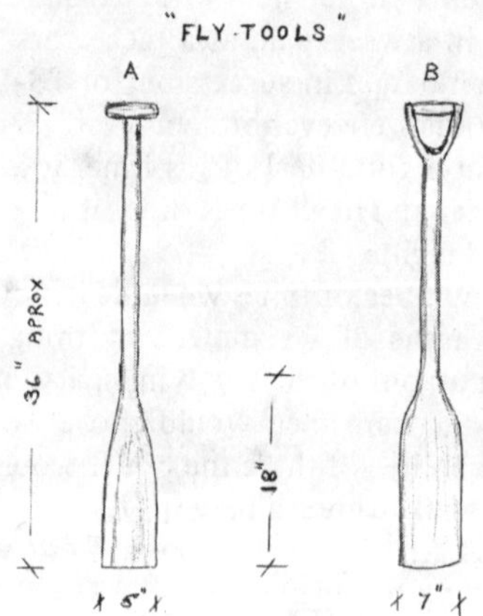

A USED FOR DIGGING AND LOADING CLAY.

B USED FOR TIDYING UP THE SLUFF OR DROPPINGS.

BOTH TYPES OF TOOL WERE MADE FROM TIMBER OF EITHER APPLE OR BEECH.

FLY-TOOLS

The main tool used by the muddies for hand digging the clay, was what was known as a fly-tool. This was, in fact, a small wooden spade made from apple or beech, three feet long with a blade measuring five inches by eighteen, which was tipped with iron.

With this implement they were able to cut square pieces of clay and flick them into the air and into a barge's hold with clockwork repetition.

Another type of fly-tool used by muddies, was a little different from the first in as much as the leading edge was half-mooned shaped. This implement was used to clear up the 'sluff' or loose wet droppings which could make a slippery berth.

The third hand implement used by the muddies, was an ordinary shovel which was used to remove the top layer of marsh peat and rotting marine vegetation. This digging away of the peat was known as 'uncallowing' and was very necessary to remove all foreign matter which could be detrimental to the manufacturing process of the cement.

CLAY CONTRACTORS

Arthur Smith, Grange Farm, Gillingham, Kent.

Before the 1900s, large parcels of land on the Stoke marshes were owned by farmers and were used mainly for grazing. Several farmers leased parts of their land to cement companies for clay digging.

One such man was Mr Arthur Smith of Gillingham, Kent, who leased a part of his Stoke marsh to the firm of John Bazley White of Swanscombe who, before the 1900s, had no less than five cement works in operation.

Bazley White worked a large clayhole on this land and ran the clay to these works.

The barges *Victory, Victoria* and the *Violet* ran clay to The Beaver and the Bridge at Frindsbury, while the *Moorhen, Plover* and *Dabchick* ran clay to the Swanscombe works.

Two of Arthur Smith's other barges *William & Jane* and *The Earle of Macclesfield* ran clay into Bazley White's Gillingham works. In 1896 John Bazley White built the *Nan* for the clay-berth at the Gillingham works; she was a 'flat iron' build for cheapness and could carry 100 tons of clay in five foot of water. Arthur Smith finished after the formation of the 'Combine'.

Oliver Smith of Gillingham, Kent

Another man who did some of the claywork before the 1900s was one Mr Oliver Smith of Gillingham – he may well have been a relative of Mr Arthur Smith.

Smith had two barges regularly employed in the claywork, they were the *Grange* and the *Mary*, both these craft carried a freight of a hundred tons of clay and mainly supplied the works on the River Thames. Each barge would do three trips a fortnight. This firm of Smith ceased trading at the turn of the century.

Henry Woolley, Stoke

Another clay contractor supplying clay to the 'Combine' at the turn of the century was a one Henry Thomas Woolley.

Wooley worked a large clayhole in an area known as Dagnum Stoke Marsh, which is alongside the Berry Wiggins Refinery, old wooden jetty.

He employed something like four gangs of muddies to dig and load the clay.

Many barges would call at this clayhole to pick up a freight rather than sail light to the cement works of either the B.B.C.L. Peter's or Lee's.

HMS Bulwark sabotaged in 1914, scattering huge fragments of steel and timber across the Stoke mud-holes.

An interesting incident happened in this mudhole when the battleship *H.M.S. Bulwark* was blown up in 1914. A huge fragment of steel weighing several hundredweight was blown over into the clayhole and crashed down through the ceiling of the barge *Dee*. Henry Woolley owned two barges of his own which did mainly general cargo, but also did a bit of claywork – they were the *Security* and *Buckland*.

Henry Woolley ceased trading before the 1920s.

T. & C. Tuff of Borstal, Kent

Another of the clay contractors supplying large quantities of clay to different cement works on the Thames and Medway was brothers Thomas & Charles Tuff. The two brothers worked several large clay-holes on the saltings at Middle Stoke.

This clay was all dug and loaded into barges by hand. Most of Tuffs

Ex-Tuff's *Gerald* hulked at Coles Wharf, Cowes, Isle of Wight 1972. Her wide decks made her ideal for clay work.

They were namely *Bertie, Mystery, Jessie, Trilby, Gerald, Caprice, Madrali, Emma Mizzen* and *Kardamah*. In the 1920s the firm of Tuff amalgamated with Walter Miskin and which also incorporated the firm of Samuel West of Gravesend. West's were one of the top barge owners at this time, and owned a large fleet of barges. Most of West's craft were bought cheap from other owners and some were of very large capacity, fit for coasting work and were very good barges, but many were old tear-outs and finished their days in the clay traffic. The firm of Tuff & Miskin ceased trading before the 1930s.

F.T. Everard of Greenhithe
Around 1924 the firm of F.T. Everard secured a contract to supply clay to 'Tunnel' cement works at West Thurrock in Essex.

And so the firm of Everards built two big pontoons at their ship-building yard at Greenhithe, Kent. On these pontoons they put Grafton steam cranes with grab buckets. One of the cranes was held in reserve at the yard whilst the other was towed down to a position in Higham Bight to dig and load clay into the firm's barges.
muddies lived in Middle Stoke – the firm had a row of terraced houses in the main street named 'Tuffs Cottages'. Most of the clay was loaded away in the firm's own barges, most of which were of a large capacity, had a little side, wide decks and being of shallow draught made them very suitable for clay work.

The driver of this crane for many years was a man named Edward (Teddie) Fowler, known affectionately to the bargemen as 'Tooting Teddy' and he was assisted by his son young Edward (Spider) Fowler.

Teddy Fowler was very fast in loading a barge, and through all his moaning and groaning could load it in about one and a half to two hours. It had been known on occasions that when he was upset or did not get any tea from the bargemen, he would slop a grabfull across and sometimes down a barge's scuttle hatch.

He once held up the stern of the *John Bayly* to stop her from sinking.

At this time Everards had up to fourteen barges in the claywork, many of them running to the Tunnel works *Jane, Mabel, Lina, Lady Maud, Lady Margery, Nellie Mary, Queen Bee, Joseph Augustas, Harriet Howard* and the *John Bayly* were but ten.

Being such a short run from Higham Bight up to West Thurrock some barges went down on a Monday morning and were back again Monday night unloaded. Whilst the barges were being unloaded by grab, they would fill up the fresh water tank for the crane, and stock up with provisions for a quick turn-round. Some barges were known to have sailed seven days a week and would get in four or five freights a week, they spent every night in Sea Reach coming or going and the crews made very good money.

Everard's did this work up until the 1930s when one would suppose their contract expired but the end came so suddenly that it was reckoned the crews of the barges nearly cried because they were doing so well out of it and it had all now come to an end.

The barges doing this work were now absolutely worn out, their timbers were all worn and the bolts rusted thin after spending a lifetime in the salt water. When four or five of them were laid up at the 'White Hart' Draw Dock at Greenhithe to break up, it was said that they went onto the sloping mud and the tide went out, they all gave a loud sigh of relief that could be heard all over Greenhithe.

BARGE SUB-CONTRACTORS

One of the small bargeing sub-contractors to the cement industry was that of Mr Arthur Larraman of Rochester.

Larraman owned a few barges that were in a good sea-worthy condition and could ply between the two rivers.

He hired his barges out to the Gillingham Portland Cement works on a slightly cheaper rate than most. His supplying of clay was always a penny a ton cheaper.

Although Arthur Larraman owned his own barges they all carried the 'Bulldog Brand' and the traditional Bulldog in their sails.

He also owned an ex-naval craft called the *Larrie* which was built almost entirely of teak and was copper fastened.

Arthur Larraman's *Fraternity* displaying the Bulldog in her sails.

This vessel traded between Gillingham and London with whiting. On the return trip would pick up a freight of coke from the Beckton Gas Works.

This vessel was once a house-boat in Conyer Creek and was broken up without any known trace. Arthur Larraman ceased operating his barges when the Gillingham factory closed in 1939.

This is a list of barges owned by Mr Arthur Larraman of Rochester.

Name	*Reg. Ton.*	*Official No.*	*Port of Registry*	*Where built*	*Year*
Dart	39	110957	Rochester	Faversham	1899
Fraternity	50	113714	Rochester	Rochester	1902
Lizzie	34	95451	London	Millwall	1882
Premier	42	112711	London	Faversham	1900

Note:– The *Larrie* had been purchased at a War Department sale and until she was sold was named *Sir Stafford Northcote*. She was built at Wivenhoe in 1889 and was stationed at Devonport in 1904.

Soon after she was purchased a 52 bhp engine was fitted and her sailes removed.

Her heavy construction made her cargo deadweight much less than similar craft.

Larrie was first registered at London in 1935. Her official number was 164462, net registered ton 44.

BARGES IN THE CLAY WORK OWNED BY SOLAMON JOHN BRICE & SON, RAINHAM & HOO, KENT

Barge	*Reg. Tons.*	*Tons of Clay Carried*	*Captain*	*Where built*	*Year*
Ada Mary	41	80	?	Frindsbury	1865
Alumina	60	115	A. Batty	Frindsbury	1899
Ash	49	115	W. Squires	Rochester	1907
Bessie Hart	42	100	D. Philpott	Sittingbourne	1866
Blue Bell	46	100	F. Webb	Milton	1862
Busy Body	49	155	Jim Batty	Rochester	1905
Conyer	37	90	?	Sittingbourne	1866
Coombdale	65	155	B. Manning	Greenwich	1887
Ella & Norman	64	145	H. Day	Strood	1885
Farola	49	125	J. Briggs	Rochester	1894
George & Ann	43	100	C. Morey	Sittingbourne	1862
George & William	41	90	E. Fiest	Sittingbourne	1879
Henry Wood	38	90	A. Webb	Sittingbourne	1877
London Belle	50	130	H. Aitkins	Limehouse	1891
Metropolis	46	110	F. Walker	Rochester	1863
Milton	38	80	?	Sittingbourne	1862
Miranda	45	110	H. Hobart	Rochester	1883
Mosquito	40	80	J. Webb	Frindsbury	1875
Nelson	63	150	H. Manning	Sittingbourne	1905
New John	38	80	?	Brentford	1878
Northfleet	29	60	?	Bankside	1862
Partridge	45	135	T. Morgan	Rochester	1887
Pioneer	43	100	A. Webb	Sittingbourne	1862
Plover	62	145	W. Sutton	Swanscombe	1898
R.S. Jackson	48	135	L. Squires	Upnor	1895
Rowland	65	150	B. Squires	Ipswich	1895
Richmond	49	110	J. Harrison	Wandsworth	1876
S.J.B.	46	110	A. Manser	Milton	1888
G.C.B.	63	160	J. Squires	Rochester	1907
San Toy	61	155	T. Squires	Rochester	1899
Sophy	38	80	?	Sittingbourne	1861
Silica	54	135	H. Wickenden	Sittingbourne	1899
Sphere	37	80	?	Rochester	1889
Thomas Harrison	41	80	?	Upnor	1866
William Bennett	42	80	F. Double	Frindsbury	1875

The *Sphere* built with no forehold, especially for the mudwork. In Deptford Creek 1925.

Miranda hulked at Damhead Creek 1963.

Miranda "shipping 'em green".

Captain Wilfrid (Pip) Box, the very last combine skipper making a come-back aboard *May* 1970.

Part VI: Barge Histories

THE END OF AN ERA

In conclusion, with the co-operation of the Society for Spritsail Barge Research we were able to trace the location of the hulks of over fifty craft that were part of the 'Combine' fleet.

Starting in the Medway we have as house-barges listed with their Lloyds official number *Dart* 110957, *MN* 99921, *Swale* 104932, *Hibernia* 110967 and *Overcomer* 109929. *Alice* 76590 buried at Halling.

Sphere 94580 and *Cygnet* 87118 buried at Rochester Esplanade. Derelict at Temple Marsh is *Russell* 78995.

Derelict at Whitewall Creek are *Rathmona* 106529, *Holborough* 81893, *Plover* 110026, *Duplicate* 110020, and *Osprey* 56751. Hulked at Strood are *Ninety* 97730 and *Arial* 67081. Hulked at Kettles Hard, Chatham is the *Ethel Margetts* 76617. Buried at Upnor is the *Wouldham Court* 109928. Hulked on the Stoke Saltings are the *Henry Tinnoth* 55141, *Swale* 47955 and the *Silica* 110961. Hulked in Stangate Creek is the *Eva* 104311.

'Combine' barges hulked in the Swale are: Hulked at Elmley the *Georgiana* 84416. Hulked at Harty Ferry are the *Lizard* 99023, and the *New World* 76624.

At the head of Conyer Creek are the remains of the *June* 58497, *J.M.W.* 84417 and the *Gore Court* 84430.

Broken up in Oare Creek was the *First Attempt* 108339. At Adelaide Dock in Milton Creek are the remains of the *Loualf* 109924 and the *Alliance* 67095. Broken up at Ramsgate Harbour was the *Marconi* 114679.

On the Thames there were several ex-Combine barges that were in use as house-boats.

Some were broken up in the Thames:

Broken up at Northfleet were the *Garfield* 84440 and the *Dunstable* 98813. Broken up at Erith were the *Edwin* 118203 and the *Hilda* 87054.

At Chiswick as house-boats are *Grace* 98791 and the *Humber* 108512. *December* 78520 was broken up there. At Kingston-on-Thames as a house-boat is the *Alumina* 110962.

A few finished their days on the Essex side and beyond. In Smallgains Creek are the hulks of *Burton* 81867, *Teal* 91890 and *Mahatma* 99008. Hulked on Slow Creek is *Cerf* 90988. *Hawk* is hulked at Maylandsea, and *Victoria* 108233 is hulked on the banks of the river Stour. *Jane Mead* 90981 is a hulk at Littlehampton, *Heron* 110174 is a houseboat at Falmouth, *Louisa* 87121 is a hulk at Cowes on the Isle-of-Wight. *Ada Mary* 94554 is a hulk at Newhaven.

That concludes the list of the 'Combine' craft. One could wonder at the whereabouts of the other two hundred and fifty superb examples of workmanship, all either sunk, broken up or burnt without trace.

If one could turn the clock back a century, one would see some strange looking craft trading in the Thames and Medway. But each craft constructed as a one-off job from bargeyards using antiquated rule of thumb or trial and error methods, like suspending a long piece of thin anchor chain from rudder post to stem post, the tightness and height would give the depth of side and a lovely sheer.

The early craft were flared sided, swim-headed and budget sterned, then they progressed to round bows, some had bluff bows, a beautiful sheer and were tiller steered. Some were converted lighters, some were of a flatiron build for cheapness and shallow draught, some were built for a particular berth, but all of them had a character all their very own.

Many were an improvement on the last one built, the criterion being to build a barge with a long serviceability, good cargo capacity, handiness, and not forgetting a fair turn of speed to windward.

Each craft could have been an epitaph to its builders, each was a complete masterpiece in collective workmanship of blacksmithery, shipwrighting, and sailmaking, with every man and boy taking a pride in his job from the laying down of the keel to the final rigging out.

Ours is the loss because most of the practical knowledge died with the men who had gained it, so little of it was passed on or even recorded.

The same thing applies to the old barge captains, grandfathers, fathers, brothers, uncles and sons who had spent a lifetime on the water and have now all passed on, not leaving behind any record of their self-taught knowledge of the art of being self-sufficient, and being able to assess the tide and wind conditions of the many rivers, creeks and navigatable stretches of water from Devon right round to Suffolk, not to mention the handling of their barges and the brilliant feats of seamanship that were accomplished and were accepted as part of their everyday working lives. Some of the old skippers knew just about every berth, hard, and wharf a barge could put into.

BARGE HISTORIES

The following factual histories of the A.P.C.M. barges came from three sources. One came from my own fleet lists of barge owners, the second one came from maritime historian Mr J. Hines who kindly checked his own records together with Lloyds shipping lists, and the third came from the actual hand-kept records of the fleet that was written in a large leather bound ledger by a senior clerk employed by the Amalgamated Portland Cement Manufacturers.

With this information at hand, we were able to give a fairly accurate history of each of the two hundred and ninety one barges that made up the 'Combine' fleet which came about by the merging and amalgamation of well over forty cement manufacturers that were to form the A.P.C.M. The 'Combine' as we know it, was officially formed on the 10th of July 1900. A year later in 1901 many other small cement companies had been bought out and their works closed down.

The 'Combine' was then known as the B.P.C.M. or the British Portland Cement Manufacturers.

In 1912 the giant firm of G.T. Earle was finally absorbed into the B.P.C.M.

Firms like Smeed Deans of Sittingbourne, Kent, G.T. Earle of Rochester and E.J. & W. Goldsmith of Rainham, Kent, that carried on independantly for a few years formed what was known as the 'Red Triangle' group.

In 1931 the 'Red Triangle' joined the B.P.C.M. – it then became known officially as the A.P.C.M. as we know it today.

During this thirty year period, well over three hundred spritsail rigged barges came under the ownership of the 'Combine'. Some of them had been disposed of by their original owners; many of them were now old and un-seaworthy and were subsequently condemned, burnt or broken up.

Others were sold at prices which were influenced by the soundness of their hulls and perhaps the fitness of their gear.

In fact they were sold from anything from one to seven hundred and fifty pounds.

Many were constructed from the very best seasoned heart timbers and had many more years of working life left in them and together with the thirty one or more craft that were rebuilt by Smeed Deans of Murston, gave very good service to their new owners. Quite a few had their gear taken out and were made into motor barges, others became lighters or 'roads' barges, some were used as yachts and many became houseboats moored permanently in Creeks, wharves and docks.

From the original number of craft in the fleet, only about fifty can be traced, the rest have faded away.

Active Official No. 63554.
Her registered tonnage was 40 reg. tons and she carried 80 tons of cement. She was built and owned by Thomas Bevan at Northfleet in 1869. Her dimensions were 72′6″ x 14′5″ x 4′9″. Was registered in London. Was sold to Herbert S. Gazalee & Sons in 1917 for £125. No known trace.

Adelaide Official No. 74812
Her registered tonnage was 48 reg. tons and she carried 90 tons of cement. She was built by E.G. Watson at Rochester in 1876 and owned and worked by Robins & Co. of Northfleet, then the A.P.C.M. Her dimensions were 77′8″ x 17′4″ x 5′7″. P.L.A. No. 10447, registered in Rochester. She was sold to Braithwaite & Co. in September 1933 for £20. They owned her till 1947. Became a hulk.

A.K. Official No. 77051
Her registered tonnage was 53 reg. tons and she carried 70 tons of cement. She was built by Alfred H. Keep at Greenhithe in 1876 and owned by John Bazley White of Swanscombe. No record of her dimensions. Was registered in London. Was sold to F.T. Everard in 1901 and was later sunk off Eastbourne. Insurance paid £594.

Alice Official No. 21885
Her registered tonnage was 50 reg. tons and she carried 90 tons of cement. She was built by J.M. Goldfinch at Faversham in 1859, launched in January, and was owned by Phillip Hilton of Faversham, then the A.P.C.M. Her dimensions were 74′7″ x 18′4″ x 6′9″. P.L.A. No. 13411, registered at Faversham. Sold to Humphrey & Grey Ltd. in February 1931 for £40.

Alice Official No. 76590
Her registered tonnage was 42 tons and she carried 75 tons of cement. She was built by William Highams at Rochester in 1877 and owned by Henry Peters of Wouldham, Kent, then B.P.C.M. No record of her dimensions, she had wooden mastcase and was stumpy rigged. P.L.A. No. 14455, was registered at Rochester. Sold to Crown & Quarry Works in August 1929. Her remains are at Whorns Place, Halling, Kent.

Alice & Ella Official No. 86612
Her registered tonnage was 45 reg. tons and she carried 110 tons of cement. She was built by D. Stocker of Wandsworth in 1882 and was owned by Geo. Roberts of Ipswich and then A.P.C.M. P.L.A. No. 13433, registered in Ipswich. She was derelict at Northfleet in 1935. She spent her latter years with the A.P.C.M. as a clay barge. Broken up at Rayfields Yard, Northfleet, 1934.

Alliance Official No. 67095
Her registered tonnage was 50 reg. tons and she carried 115 tons of cement. She was built by E.G. Watson at Rochester in 1874 and was

owned by Robins & Co. London and A.P.C.M. Her dimensions were 79′0″ x 18′10″ x 6′0″. P.L.A. No. 10448, registered at Rochester. She was sold to A.G. Wood of Sittingbourne in 1935 for £2. Her hulk is believed to be lying in Milton Creek, Sittingbourne. Her latter years with the 'Combine' were spent in the clay work, having her holds lined with steel tanks.

Ex-Gibbs & Co *Alliance* hulked at Murston 1960.

Alliance II Official No. 85083
Her registered tonnage was 43 tons and she carried 105 tons of cement. She was built by F. White at Brentford in 1881 and was owned by Gibbs & Co. West Thurrock and A.P.C.M. Her dimensions were 75′11″ x 17′9″ x 6′2″. P.L.A. No. 6945 and registered at London. She was tied up at the Sittingbourne Ash Traffic. Her latter days with the 'Combine' were spent in the clay work.

Ally Sloper Official No. 97724
Her registered tonnage was 60 reg. tons and she carried 115 tons of cement. She was built by William Highams at Rochester in 1890 and launched in July of that year. She was owned by Martin-Earles of Strood, Kent and A.P.C.M. Her dimensions were 76′0″ x 18′6″ x 5′9″. She was registered at Rochester, a stumpy barge. She was sold to H. Lea in November 1932 for £50. Took her name from a comic character.

Alumina Official No. 110962
Her registered tonnage was 60 reg. tons and she carried 140 tons of cement. She was built by J. Gill & Son at Rochester in 1899 and owned by Henry Peters of Wouldham & Burham, then B.P.C.M. Her

dimensions were 79′6″ x 21′6″ x 5′9″. P.L.A. No. 14456, she was registered in Rochester. She was sold to S.J. Brice in June 1928 and worked in the clay work till about 1944. Resold. Converted to a Motor Barge owned by Sheppey Motor Barges. Built and worked as a clay barge, worked to the Burham Works. Was a house-boat at Kingston-on-Thames. 1948.

Ambrose Official No. 20571
Her registered tonnage was 41 reg. tons and she carried 80 tons of cement. She was built at Frindsbury in 1858 and owned by Thomas Weekes of Halling, Trenchman Weekes & Co., Halling and A.P.C.M. Her dimensions were 72′8″ x 14′4″ x 5′5″. Registered at Rochester. Was sold to Messrs. W.J. Beamont in 1915 for £20. No known trace.

Amy Seymour Official No. 78503
Her registered tonnage was 64 reg. tons and she carried 110 tons of cement. Was built by J. Seymour at Strood in 1877 and was launched in the October of that year. Was owned by John Seymour of Chatham and A.P.C.M. Her dimensions were 84′0″ x 19′7″ x 6′1″. P.L.A. No. 9113 and she was registered in Rochester. She was sold to the Dorset Quarry Co. in 1926 for £250. Took her name from the builders wife. Reported derelict at Poole 1950.

Ann Official No. 58463
Her registered tonnage was 36 reg. tons and she carried 74 tons of cement. She was built at Murston in 1868 by the B.B.C.L. Co. and was owned by The Burham Brick Cement & Lime Co., then by the A.P.C.M. She was Registered at Rochester. No known trace.

Anne Maria Official No. 50303
Her registered tonnage was 44 reg. tons and she carried 90 tons of cement. She was built by J.M. Goldfinch at Faversham in 1864 and was owned by Philip Hilton of Upnor, then A.P.C.M. Her dimensions were 73′6″ x 17′0″ x 5′10″. P.L.A. No. 13421 and she was registered at Faversham. She was sold to Mr Bellingham in 1930 for £80. Her hulk lies in Higham Bight.

Annie Official No. 50171
Her registered tonnage was 42 reg. tons and she carried 85 tons of cement. She was built at Pangebourne in 1864 and was owned by Josiah Jackson of Shoebury. She was registered in London. Came under the A.P.C.M. ownership. No known trace.

Answers Official No. 104312
Her registered tonnage was 45 reg. tons and she carried 115 tons of cement. She was built by William Higham at Rochester in 1894 and was owned by Martin Earle & Co., then A.P.C.M. Her dimensions were 78′4″ x 14′9″ x 5′5″. P.L.A. No. 11246, registered at Rochester. She was sold to R.H. Penny & Sons in 1931 for £220. Converted to a motor barge 1932.

Ant Official No. 118264
Her registered tonnage was 69 reg. tons and she carried 130 tons of cement. Her P.L.A. No. 9001. She was owned by A.P.C.M. Was swim headed and was given away with £5. A bit of a mystery this barge, could have been a converted lighter. No known trace.

April Official No. 58486
Her registered tonnage was 41 reg. tons and she carried 85 tons of cement. She was built at Aylesford in 1869 by the B.B.L.C. for their own use, then A.P.C.M. P.L.A. No. 10504, was registered in Rochester. Was sold in March 1926 to Smeed and Wadhams for £30, used as a coal hulk. Buried in the old Gillingham Gas Works.

Ariel Official No. 67081
Her registered tonnage was 37 reg. tons and she carried 75 tons of cement. She was built in 1874 at Rochester and was launched in June of that year. She was owned by Samuel Smith, Halling, William Lee, B.P.C.M. and A.P.C.M. Her dimensions were 74′0″ x 15′1″ x 5′6″. P.L.A. No. 14457 and was registered at Rochester. Was sold to R.W. Austin in June 1934 for £5, then Wadhams & Smeed. Was derelict at Strood.

Arthur
No other details except that she was sold to The Acorn Barge Co. in 1901 for £60.

Arthur Margetts Official No. 90979
Her registered tonnage was 34 reg. tons and she carried 105 tons of cement. She was built at Aylesford in 1885 by The West Kent Gault Brick & Cement Co. and launched in November of that year. Was owned by the builders, then A.P.C.M. Her dimensions were 80′6″ x 17′9″ x 5′10″. P.L.A. No. 14458, she was registered in Rochester. Was sold to Mr F. Butler of London in October 1931 for £120. Became a yacht barge in 1937. Hulked somewhere on the River Yare, Norfolk.

August Official No. 67058
Her registered tonnage was 66 reg. tons. She was built at Aylesford in 1873 for the Burham Brick, Cement & Lime Co. Posted missing 30th March 1883. Was bound from Rochester to Hull with cement 19th January 1883. Wind was S.E. and flew to N.W. gale.

August Official No. 90979
Her registered tonnage was 44 reg. tons and she carried 90 tons of cement. She was built by The B.B.C.L. Co. at Aylesford in 1886, was launched in May that year. She was owned by this company at Burham, then the A.P.C.M. Her dimensions were 77′6″ x 17′0″ x 5′9″. P.L.A. No. 10508 and she was registered in Rochester. Was sold to Mr R.C. Foreman in April 1933 for £15. Derelict near East Greenwich 1947.

Barges in the Swale. In the centre is an artist's impression of the mud-barge 'Admiral Blake.'

Autumn Official No. 84378
Her registered tonnage was 43 reg. tons and she carried 90 tons of cement. She was built by B.B.C.L. Co. at Aylesford in 1881 and owned by the Burham Brick Lime and Cement Co., Burham, then A.P.C.M. Her dimensions were 72′8″ x 14′6″ x 4′9″. She was registered in Rochester. She was condemned and broken up.

Beagle Official No. 81894
Her registered tonnage was 42 reg. tons and she carried 80 tons of cement. She was built by J. Gill & Sons at Rochester in 1880 and was owned by Henry Peters of Burham, then B.P.C.M. Her dimensions were 73′9″ x 15′4″ x 5′6″ and was stumpie rigged and tiller steered. P.L.A. No. 14459, was registered at Rochester. She was sold in September 1929 to Kingsnorth Works Dept. for £30 and was resold to W.H. Shepherd in September 1933. No known trace.

Bee Official No. 118267
Her registered tonnage was 86 reg. tons and she carried 180 tons of cement. Her owners were A.P.C.M. Her dimensions were 79′2″ x 21′3″ x 7′1″. Swimheaded. P.L.A. No. 9003, registered in London. Sold in June 1932 to Bradbury Son & Co. for £45. Could have been lighter, rigged. No known trace.

Bella Official No. 74808
Her registered tonnage was 35 reg. tons and she carried 65 tons of cement. She was built by Geo. Curel at Frindsbury in 1876 and was launched in May of that year. Was owned by Thomas Weekes-Trenchman Weekes and B.P.C.M. P.L.A. No. 14461, registered in Rochester. Was condemned and sold to A. Hart in December 1926. She was stumpie rigged. No known trace.

Black Duck Official No. 86983.
Her registered tonnage was 58 reg. tons and she carried 105 tons of cement. She was built by John Bazley Whites at Swanscombe in 1882 and launched in August of that year. Was owned by this company then by A.P.C.M. Her dimensions were 82′6″ x 19′0″ x 5′9″. P.L.A. No. 8974, was registered in London. Her hull was sold in June 1932 to The Thames Transport Co. for £40. She was named after the place where she was built, Black Duck Wharf. No known trace.

Blackfriars Official No. 98800
Her registered tonnage was 42 reg. tons and she carried 80 tons of cement. She was built and registered in Rochester in 1891, and was owned by Samuel Smith, Halling, E. Wall, London and B.P.C.M. P.L.A. No. 14460. Was sold to The Braithwaite Engineering Co. on February 20th 1933 (hull only). Derelict at Conyer 1938.

Bride Official No. 89678
Her registered tonnage was 42 reg. tons and she carried 95 tons of

cement. She was built at Sittingbourne by R.M. Shrubsall in 1885 and owned by A.P.C.M. Her dimensions were 78′6″ x 17′3″ x 5′9″. P.L.A. No. 10449, registered in London. Was sold to A.G. Wood in February 1937 for £5. Was a lighter in 1939. Derelict at Murston in 1945.

Brisk Official No. 65738
Her registered tonnage was 41 reg. tons and she carried 80 tons of cement. She was built at Northfleet in 1872 and was owned by Thomas Bevan of Northfleet, then A.P.C.M. Her dimensions were 72′7″ x 14′7″ x 4′6″. Registered in London. Was sold in 1918 to C. Smith for £180. No known trace.

Ex-West Kent's *Britannia* circa 1933.

Brittania Official No. 113682
Her registered tonnage was 46 reg. tons and she carried 120 tons of cement. She was built at Rochester by Gill & Son in 1900 and was owned by The West Kent Portland Cement Co., then the B.P.C.M. Her dimensions were 79′0″ x 17′8″ x 6′3″. P.L.A. No. 14461, registered at Rochester. Was sold to Tester Bros. in December 1934 for £125. Was a hulk in 1949.

Cader Idris Official No. 90969
Her registered tonnage was 50 reg. tons and she carried 125 tons of

cement. She was built at Sittingbourne in 1884 by S. & J. Taylor and was launched in the November of that year. She was owned by E.J. Brooks of Grays then the A.P.C.M. Her dimensions were 82′0″ x 18′4″ x 6′2″. P.L.A. No. 13404, she was registered at Rochester. She sank in a gale on 12.1.30 in Sea Reach. Southend Life Boat rescued the crew and her hull was sold to S. Bev of Leigh.

Caledonia Official No. 110959
Her registered tonnage was 46 reg. tons and she carried 115 tons of cement. She was built at Rochester in 1899 by Gill & Sons and was launched in July of that year. She was owned by The West Kent Portland Cement Co. Burham then B.P.C.M. Her dimensions were 78′6″ x 17′10″ x 6′3″. P.L.A. No. 14462, Registered at Rochester. She was sold to T. Watson of Rochester in May 1928 for £450. In 1933 sold to L. Surridge – hulked in 1940. No known trace.

Cecil Margetts Official No. 97718
Her registered tonnage was 46 reg. tons and she carried 115 tons of cement. She was built at Frindsbury by Geo. Curel in 1890 and was launched in the April of that year. She was owned by The West Kent Portland Cement Co., Burham, Kent, then B.P.C.M. Her dimensions were 77′6″ x 18′2″ x 6′4″. P.L.A. No. 14463, registered at Rochester. She was sold to William Cory & Sons Ltd in October 1930, then owned by The Stuart Sand Co. 1933. Taken over in 1941 by H.M. Government as a mine spotting hulk.

Ex-Phoenix Portland Cement Company's *Cerf* hulked at Stow Creek Essex 1967

Cerf Official No. 90988
Her registered tonnage was 44 reg. tons and she carried 110 tons of cement. She was built by Geo. H. Curel at Frindsbury in 1886 and was launched in the June of that year. She was owned by The Phoenix Cement Co. of Frindsbury, then A.P.C.M. Her dimensions were 76′6″ x 17′8″ x 6′5″. P.L.A. No. 13440, Registered at Rochester. She was sold in May 1934 to R. Stuart for £100. Her hulk lies in Slow Creek on The River Crouch.

Challenge Official No. 70723
Her registered tonnage was 42 reg. tons and she carried 90 tons of cement. She was built at Upchurch by R.M. Shrubsall in 1875 and owned by Thomas E. Wood of Gravesend then A.P.C.M. She was sold in June 1932 to A.C.E. Jemmett for £10.10s., then owned by Sam Elliot of Milton, Sittingbourne. No known trace.

Charles Official No. 12590
Her registered tonnage was 30 reg. tons and she carried 60 tons of cement. She was built at Brentford in 1854 and owned by William Tingey of Frindsbury, then A.P.C.M. Her dimensions were 70′1″ x 14′1″ x 3′9″. She was registered in Rochester. A very early craft – she was condemned. Broken up without trace.

Charlotte Official No. 4576
Her registered tonnage was 46 reg. tons and she carried 75 tons of cement. She was built at Strood in 1847 and was owned by William Tingey of Frindsbury, then the A.P.C.M. Her dimensions were 72′7″ x 14′5″ x 4′8″. Registered at Rochester. Another very early craft – condemned.

Clenwood Official No. 127256
Her registered tonnage was 61 reg. tons and she carried 160 tons of cement. She was built at Borstal in 1911 by James Little, owned by this company, then the A.P.C.M. Her dimensions were 84′1″ x 19′5″ x 6′4″. Registered at Rochester. She was sold to T.A. Whiting, converted to a Motor Barge in 1928 and worked around the Isle of Wight. Resold by Capt. Whiting. Sank near the Nore Sand.

Clipper Official No. 67104
Her registered tonnage was 42 reg. tons and she carried 90 tons of cement. She was built at Rochester in 1875 by E.G. Watson and was owned by Robins & Co. Westminster Lighteridge Co., Westminster, then A.P.C.M. Her dimensions were 77′2″ x 16′8″ x 5′0″. She was registered at Rochester – had a motor fitted in 1917. Sold to J. Rose in 1921 for £200. Towed into Portsmouth by H.M. Tug ‘St Just’ March 1923.

Clyde Official No. 87221
Her registered tonnage was 41 reg. tons and she carried 80 tons of cement. She was built at Rochester in 1883 by Gill & Son and was launched in November of that year. She was owned by G.K. Anderson of Halling, then A.P.C.M. Her dimensions were 73′0″ x 15′3″ x 5′5½″. P.L.A. No. 13420, registered at Rochester. She was sold in September 1933 to J. Mowlem & Co. for £60. She was stumpie rigged. Hulked at Upnor 1933.

Cotswold Official No. 98137
Her registered tonnage was 78 reg. tons and she carried 160 tons of

Ex-Lee's *Clyde* displaying an early A.P.C.M. in the peak of her mainsail.

cement. She was built in Blackwall by A. White in 1887. Registered in London. A big swimheaded barge (no other details) could have been a lighter. Lost without trace.

Crow Official No. 101911
Her registered tonnage was 60 reg. tons and she carried 125 tons of cement. She was built at Northfleet in 1891 by Thomas Bevan. She was owned by this company, then the A.P.C.M. Her dimensions were 80′6″ x 19′3″ x 6′6″. P.L.A. No. 9753, registered in London. Sold in 1931 to H.R. Mitchell & Sons for £175. Broken up in 1948 at Gravesend.

Cullen Official No. 84393
Her registered tonnage was 42 reg. tons and she carried 85 tons of

cement. She was built by Geo. H. Currel of Frindsbury in 1894, owned by this company then the A.P.C.M. She was registered in Rochester. No known trace.

Cygnet Official No. 87118
Her registered tonnage was 57 reg. tons and she carried 140 tons of cement. She was built at Swanscombe by John Bazley White in 1883, probably a 'Swimmie'. She was owned by J.B. White & Co. of Swanscombe, then the A.P.C.M. She was registered in London. Was a hulk in 1933.

Dabchick Official No. 89670
Her registered tonnage was 62 reg. tons and she carried 130 tons of cement. She was built by John Bazley White at Swanscombe in 1885 who also owned her. Her dimensions were 84′0″ x 19′2″ x 5′9″. P.L.A. No. 8996 registered in London. She was sold to W.H. Hills in May 1931 for £250. Laid up at Strood 1938.

Daisy Official No. 52905
Her registered tonnage was 42 reg. tons and she carried 80 tons of cement. She was built by Mark Reaman of Faversham in 1886 and was launched in the September of that year. She was owned by Anderson of Halling, then H.E. Coulter of Faversham, then A.P.C.M. Her dimensions were 72′5″ x 16′8″ x 4′8″. She was sold in June 1928 to Wadhams & Smeed for £17.10s. Buried in the old Gillingham Gas Works.

Dart Official No. 110957
Her registered tonnage was 30 reg. tons and she carried 100 tons of cement. She was built at Faversham in 1899 and was owned by Hilton Anderson of Halling, then A.P.C.M. Her dimensions were 76′4″ x 17′5″ x 5′9″. P.L.A. No. 13428, registered at Rochester. She was sold in September 1926 to Arthur Larramen for £500. Laid up in 1939. Was a houseboat in 1949.

Dart Official No. 84418
Her registered tonnage was 43 reg. tons and she carried 85 tons of cement. She was built at Cowes in 1881 and was owned by Samuel Smith of Halling, then William Lee & Sons, Halling, then A.P.C.M. She was registered at Rochester. Condemned – broken up without trace.

December Official No. 78520
Her registered tonnage was 43 reg. tons and she carried 85 tons of cement. She was built by The Burham Brick & Lime & Cement Co. at Burham in 1878. She was owned by this company then the A.P.C.M. Her dimensions were 78′6″ x 17′6″ x 5′8″. P.L.A. No. 10512, she was registered in Rochester. Laid up in 1928. She was sold to C. Bow in May 1930 for £7.10s. Was broken up at Chiswick in 1957.

Dee Official No. 109914
Her registered tonnage was 46 reg. tons and she carried 110 tons of cement. She was built at Sittingbourne by Alfred White in 1898 and owned by Hilton Anderson of Halling, then A.P.C.M. Her dimensions were 78′6″ x 17′9″ x 5′11″. P.L.A. No. 13417. Registered at Rochester. She was sold to T. Allsworth of Queenborough in April 1931 for £225. Converted to a Motor Barge in 1933. Had a 15 hp engine. Was a houseboat at Borstal 1976.

Diamond Official No. 108515
Her registered tonnage was 44 tons and she carried 75 tons of cement. She was built at Rochester by F. Sollet in 1897 and launched in December of that year. She was owned by William Tingey of Rochester then the A.P.C.M. Her dimensions were 71′0″ x 17′3″ x 4′11″. She was registered at Rochester. She was sold to S.W. Downing of Rotherhithe in February 1928 for £120. Lost without trace.

Dolphin Official No. 84386
Her registered tonnage was 45 reg. tons and she carried 80 tons of cement. She was built at East Cowes by J.S. White in 1881 and was launched in the March of that year. She was owned by Samuel Smith of Halling, then William Lee of Halling, then A.P.C.M. Her dimensions were 75′0″ x 14′10″ x 4′11″. P.L.A. No. 14465, she was registered at Rochester. She was sold to G. Mann in September 1926 for £5. Lost without trace.

Dorothy Official No. 110116
Her registered tonnage was 49 reg. tons and she carried 120 tons of cement. She was built at Limehouse by H. Shrubsall in 1899 and was launched in May of that year. She was owned by H. Wills of Gravesend, The Tolhurst Cement Co., B.P.C.M., and A.P.C.M. Her dimensions were 80′0″ x 17′6″ x 5′6″. P.L.A. No. 11365, registered at Rochester. Sold in October 1930 to W. Cory & Sons Ltd. for £180. In 1933 she was owned by The Stuart Sand & Gravel Co. Thamesmouth. Hulk used as a minespotter. Registry closed 1955.

Drake Official No. 110109
Her registered tonnage was 61 reg. tons and she carried 150 tons of cement. She was built by F.T. Everard at Greenhithe in 1899 and owned by Thomas Bevan of Northfleet then A.P.C.M. P.L.A. No. 9727, she was registered in London. She was sold to Thomas Scholey in July 1927 for £612 – was in collision at Upnor and sank circa 1930. Broken up at Upnor.

Duplicate Official No. 110020
Her registered tonnage was 50 reg. tons and she carried 125 tons of cement. She was built at Rochester by William Highams in 1898 and owned by The West Thurrock Lighterage Co., then A.P.C.M. Her

dimensions were 83′6″ x 20′2″ x 6′3″. P.L.A. No. 7109, registered at Rochester. She was sold to S.J. Peters of Southend in October 1928 for £663. Working on the Isle of Wight, motorised in 1940. Derelict in Whitewall Creek after 1950.

East Kent Official No. 44095
Her registered tonnage was 38 reg. tons and she carried 80 tons of cement. She was built at Sittingbourne in 1862 by S. Taylor and was owned by Thomas Weekes of Halling & Trenchman Weekes of Halling. Her dimensions were 72′4″ x 14′6″ x 4′7″. P.L.A. No. 14440, registered at Rochester. Was sold to F.E. Walker Ltd. in June 1931 for £20. Was a lighter in 1932. No known trace.

Eddystone Official No. 94579
Her registered tonnage was 50 reg. tons and she carried 115 tons of cement. She was built at Rochester by Gill & Son in 1889 and was owned by Francis & Co. of Nine Elms, then A.P.C.M. Her dimensions were 80′6″ x 18′6″ x 5′10″. P.L.A. No. 11109, registered at Rochester. She was sold to William Cory & Sons Ltd in October 1930 for £170, then the Stuart Sand & Shingle Co. She was sunk in a collision with S.S. City of Brussels on 4.12.1935. Broken up at Charlton.

Edward & Charles Official No. 28533
Her registered tonnage was 43 reg. tons and she carried 90 tons of cement. She was built at Queenborough, Isle of Sheppey by Japlet J. Page in 1860 and was owned by George Mantle of Faversham, then George West of Charlton, then A.P.C.M. Her dimensions were 71′0″ x 16′2″ x 5′0″. P.L.A. No. 9005, registered in London. Laid up in 1928. She was sold to F.E. Walker Ltd. in June 1931 for £20, for a lighter. No known trace.

Edward & William Official No. 78530
Her registered tonnage was 40 reg. tons and she carried 80 tons of cement. She was built at Frindsbury by Geo. Curel in 1878 and was owned by Trenchman Weekes, then A.P.C.M. She was registered in Rochester. Was condemned and sold to M. Hodge in May 1927 for £4. Broken up without trace.

Edwin Official No. 118203
Her registered tonnage was 40 tons and she carried 80 tons of cement. She was built at Sittingbourne by Eastwood & Co. in 1903 and was owned by Henry Peters of Wouldham, then A.P.C.M. Her dimensions were 76′5″ x 16′11″ x 5′3″. A stumpie barge. P.L.A. No. 14467, registered at Rochester. She was sold to C.H. Musslewhite & Sons in November 1931 for £150. Derelict near Harwich 1956

Edwin & Emily Official No. 84400
Her registered tonnage was 44 reg. tons and she carried 90 tons of cement. She was built at Conyer by J. Bird in 1881 and owned by Edwin Killick of Northfleet, then the A.P.C.M. Her dimensions were 76′6″ x 17′9″ x 6′0″. P.L.A. No. 13527, registered at Cowes. Originally at Rochester. She was sold to C. New in July 1927 for £120. No known trace.

Eliza Official No. 47941
Her registered tonnage was 41 tons and she carried 80 tons of cement. She was built at Frindsbury in 1863 and was owned by William Tingey of Frindsbury, then the A.P.C.M. Her dimensions were 73′1″ x 14′4″ x 5′1″. She was registered in Rochester. She was sold to W.T. Beaumont in 1915 for £20. No known trace.

Emma Official No. 110977
Her registered tonnage was 35 reg. tons and she carried 90 tons of cement. She was built at Milton by A. White in 1899 and owned by Henry Peters of Wouldham, the B.P.C.M. then the A.P.C.M. Her dimensions were 75′0″ x 14′9″ x 5′11″. P.L.A. No. 14468, registered in Rochester. She was sold to W.H. Theobald in April 1935 for £50. Hulked in 1949. No known trace.

Emu Official No. 112694
Her registered tonnage was 52 reg. tons and she carried 120 tons of cement. She was built at Greenhithe by F.T. Everard in 1900 and was owned by Thomas Bevan of Northfleet then the A.P.C.M. Her dimensions were 80′0″ x 18′1″ x 6′1″. P.L.A. No. 9895, registered in London. She was sold to William Cory & Sons Ltd. in October 1930 for £170, then to The Stuart Sand & Shingle Co. No known trace.

Enterprise Official No. 81860
Her registered tonnage was 41 reg. tons and she carried 80 tons of cement. She was built at Cowes, Isle of Wight by J.S. White in 1879 and was owned by Samuel Smith of Halling, The William Lee Co. Halling then A.P.C.M. Her dimensions were 74′9″ x 14′7″ x 5′8″. Composite built. P.L.A. No. 14469, registered at Rochester. She was sold to C.L. Wilson in February 1926 for £7.10s.

Estelle Official No. 110956
Her registered tonnage was 44 reg. tons and she carried 105 tons of cement. She was built at Rye by G. & T. Smith in 1899 and owned by The Formby Cement Co. of Frindsbury and Halling, The Trenchman Weekes & Co., then the B.P.C.M., A.P.C.M. Her dimensions were 81′0″ x 17′10″ x 5′9″. P.L.A. No. 14441, registered at Rochester. She was sold to G.A. Wells in February 1929 for £300. In 1934 she was then owned by Wakely Bros. of Rainham Kent. A road barge in 1947. Derelict on the sea wall near Pitsea 1957.

Ethel Margetts Official No. 7617
Her registered tonnage was 42 reg. tons and she carried 85 tons of cement. She was built at Strood by G.H. Curel in 1877 and owned by A.P. Margetts of Wouldham, Kent, then the West Kent Cement Co. and the A.P.C.M. P.L.A. No. 14470, registered at Rochester. She was sold to Wadhams & Smeed in June 1932. Yacht club hulk 1935. Hulked at Kettles Hard, Chatham. Broken up in 1975.

Eva Official No. 104311
Her registered tonnage was 50 reg. tons and she carried 120 tons of cement. She was built at Rochester by Gill & Son in 1894 and was launched in November of that year. She was owned by Geo. Clifford of Gillingham, then the A.P.C.M. P.L.A. No. 9118, registered at Rochester. She was sold to W.H. Hills in November 1926, laid up in 1938 then sold to H.M. Government. She was sunk on war service and her hulk lies in Stangate Creek on The River Medway.

Falcon Official No. 87088
Her registered tonnage was 59 reg. tons and she carried 105 tons of cement. She was built at Northfleet by Thomas Bevan in 1883 and was launched in the April of that year. She was owned by Thomas Bevan of Northfleet then A.P.C.M. Her dimensions were 76′0″ x 17′3″ x 6′1″. P.L.A. No. 9748, registered in London. She was sold to John S. Rayfield in June 1928 for £297. Derelict at Northfleet 1937.

Fanny Official No. 58509
Her Registered tonnage was 41 reg. tons and she carried 80 tons of cement. She was built at Rochester in 1870 by John Bazley White of Swanscombe. She was owned by this company, then the A.P.C.M. She was registered in Rochester. She was sold prior to 1925 to The Union Lighterage Company. No known trace.

February Official No. 58483
Her registered tonnage was 35 reg. tons and she carried 80 tons of cement. She was built at Aylesford by the Burham Brick, Lime & Cement Co. Burham in 1868, who also owned and worked her. Her dimensions were 72′0″ x 14′6″ x 4′6″. She was registered at Rochester. Her dimensions indicate that she was a canal barge. She was condemned and lost without trace.

First Attempt Official No. 108339
Her registered tonnage was 40 reg. tons and she carried 75 tons of cement. She was built at Rochester by Curel & Cory in 1898 and was owned by The West Thurrock Lighterage Co. then the A.P.C.M. Her dimensions were 74′6″ x 14′10″ x 4′7″. P.L.A. No. 9956, registered in London. She was sold to Mr T. Sempill in April 1933 for £75. She was stumpy rigged and probably a canal barge. She was a barge yacht, broken up at Oare Creek, Faversham 1958. Note – this was the first barge to be built by Curel and Cory at Rochester.

Florence Myall Official No. 90968
Her registered tonnage was 58 reg. tons and she carried 115 tons of cement. She was built at Rochester by William Highams in 1884 and was launched in the August of that year. She was owned by Walter Myall of Chatham, then William Tingey of Frindsbury, then A.P.C.M. Her dimensions were 81′0″ x 18′8″ x 6′3″. P.L.A. No. 11047, registered at Rochester. She was sold to W.H. Theobald in January 1935 for £69. Stranded on Ridge Sand on voyage from Rowhedge to Benfleet with sand. Clacton lifeboat saved crew in strong N.E. wind. Barge became a total wreck, 4th May 1938.

Flower-of-Kent Official No. 49839
Her registered tonnage was 44 reg. tons and she carried 75 tons of cement. She was built at Frindsbury in 1864 and was launched in the November of that year. She was owned by Thomas G. Simmonds of Chatham, then the A.P.C.M. Her dimensions were 75′0″ x 16′6″ x 6′0″. P.L.A. No. 13454, registered at Rochester. Laid up in June 1928. She was sold to Mr L.H. Whiley in May 1933. Hulk off Gillingham Strand.

Forth Official No. 90975
Her registered tonnage was 42 reg. tons and she carried 85 tons of cement. She was built at Rochester by Gill & Sons in 1885 and was launched in the April of that year. She was owned by George Anderson of Frindsbury, then the A.P.C.M. P.L.A. No. 13412, registered at Rochester. Sunk in collision off Erith 1924. Crew lost. She was sold to The Ministry of Pensions in May 1925 for £6.10s.

Foxhound Official No. 81895
Her registered tonnage was 42 reg. tons and she carried 80 tons of cement. She was built at Rochester by William Highams in 1880 and was launched in September of that year. She was owned by Henry Peters of Wouldham, then A.P.C.M. Her dimensions were 73′8″ x 12′5″ x 5′0″. She was registered at Rochester. She was sold to The Upnor Shipbreaking Co. in 1923 for £15. No known trace.

Friday Official No. 108520
Her registered tonnage was 46 reg. tons and she carried 115 tons of cement. She was built at Rochester by Gill & Son in 1898 and was launched in the July of that year. She was owned by the Burham Brick, Lime & Cement Co., then the A.P.C.M. Her dimensions were 78′6″ x 17′10″ x 6′2″. P.L.A. No. 10535, she was registered at Rochester. She was sold to Thomas Watson in May 1928 for £412.15s. Later sold to L. Surridge. Finished sailing about 1940. 1941 was a Mine Spotting Hulk in the Thames.

Garland Official No. 90988
Her registered tonnage was 44 reg. tons and she carried 110 tons of cement. She was built at Rochester by J. Gill & Son in 1887 and owned by William Morgan, D.C. Rutter-Wakering Brick Co., then A.P.C.M. Her

Genesta rounding the Mouse lightship in the 1929 Medway match.

dimensions were 76′9″ x 18′0″ x 5′10″. P.L.A. No. 13453, she was registered in Rochester. She was sold to The Canvey Supply Co. Ltd., in April 1931 for £150. Resold to J. Rayfield. Laid up 1939. Broken up 1948.

Genesta Official No. 118285
Her registered tonnage was 54 reg. tons and she carried 140 tons of cement. She was built at East Greenwich by H. Shrubsall in 1903 and owned by J.G. Hammond of London. Her dimensions were 84′6″ x 19′6″ x 6′0″. P.L.A. No. 13785, registered in London. She was bought by the 'Combine' in August 1906 for £885. She was a big, fast barge in her day, she excelled in many of the barge matches. 27 October 1922 sunk in collision off Rotherhithe. She was sunk in a collision off Northfleet in 1939 but was raised and resold to Whiting Bros. converted to a Motor Barge in 1942. She was sunk again in collision in the Medway in 1950 near Hoo Fort with a cargo of beer. She was burnt circa 1950 at Milton Churchfields.

George Official No. 20043
Her registered tonnage was 45 reg. tons and she carried 110 tons of cement. She was built at Frindsbury in 1855 and was owned by Thomas Weekes of Burham, Trenchmen Weekes of Burham, then the A.P.C.M. She was registered at Rochester. She was condemned and sold to Mr M. Hodges in 1922 for £5. No known trace.

George & Jane Official No. 81879
Her registered tonnage was 35 reg. tons and she carried 70 tons of cement. She was built at Millwall by R.M. Surridge in 1880 and was launched in May of that year. She was owned by J. Westwood of

Little George, a clay barge. Skipper Bill Wadhams was on a weekly wage basis.

Charlton, then the A.P.C.M. Her dimensions were 74′6″ x 15′1″ x 5′4″. P.L.A. No. 9004. She was sold to H.G. Baker of Brighton in May 1933 for £33. No known trace.

Goliath Official No. 16614
Her registered tonnage was 30 reg. tons and she carried 60 tons of cement. Her dimensions were 56′5″ x 15′3″. Probably a lighter rigged. Condemned by the A.P.C.M. No known trace.

Goshawk Official No. 89534
Her registered tonnage was 66 reg. tons and she carried 115 tons of cement. She was built at Northfleet by T. Bevan in 1884 and was launched in the March of that year. She was owned by Thomas Bevan of Northfleet then the A.P.C.M. Her dimensions were 80′6″ x 19′0″ x 6′2″. P.L.A. No. 9750 was registered in London. She was sold to Henry R. Mitchell of Charlton in April 1931 for £175. Eventually dismantled 1940s.

Gun Official No. 45511
Her registered tonnage was 41 reg. tons and she carried 80 tons of cement. She was built at Frindsbury in 1863 and owned by The Burham Brick, Lime & Cement Co., then the A.P.C.M. Her dimensions were 72′9″ x 14′4″ x 5′1″. She was registered at Rochester. She was sold to Martin Earle & Co. in 1920 for £70. A cut or canal barge. No known trace.

Gundolph Official No. 89534
Her registered tonnage was 44 reg. tons and she carried 90 tons of cement. She was built at Frindsbury by George H. Curel in 1875 and was launched in the February of that year. She was owned by J. Bouldon of Rochester, Alfred Tolhurst of Northfleet, then the

A.P.C.M. Her dimensions were 78′6″ x 17′6″ x 5′7″. P.L.A. No. 3897 registered in Rochester. Laid up in 1929. She was sold to Sageman & Phillips in January 1931 for £5. (She was built for the Landlord James Bouldon, who kept the Public House 'Gundolph' at Rochester Bridge. No known trace.

Gwendoline Official No. 84444
Her registered tonnage was 44 reg. tons and she carried 80 tons of cement. She was built at Rochester in 1882 and owned by William Tingey of Rochester, George Weedon of Milton, Sittingbourne, then the A.P.C.M. Her dimensions were 74′0″ x 14′9″ x 5′9″. She was registered in Rochester. She was condemned and sold to Mr Mann in 1926 for £7. No known trace.

Ex-Margetts *Harold Margetts* off the Upnor barge yard.

Harold Margetts Official No. 97717
Her registered tonnage was 45 reg. tons and she carried 115 tons of cement. She was built by George H. Curel at Frindsbury in 1890 and was launched in the April of that year. She was owned by The West Kent Portland Cement Co. Burham. Her dimensions were 77′5″ x 14′9″ x 5′9″. P.L.A. No. 14472, registered at Rochester. She was sold to Mr G. Andrews in October 1933 for £150. A fast barge in her day and excelled in several barge matches. Sank in 1935. Raised and sold to Gilletts (Faversham). A Barge Yacht in 1947.

Harry Official No. 52955
Her registered tonnage was 39 reg. tons and she carried 85 tons of cement. She was built at Rochester by Edmund G. Watson in 1865 and was owned by Henry Peters of Burham & Wouldham. She was registered at Rochester. She was sold to Mr. W. Chantler in 1912. 1919–22 owned by E.J. Ridgeway. No known trace.

Hawk Official No. 78509
Her registered tonnage was 43 reg. tons and she carried 80 tons of cement. She was built at Frindsbury in 1878 and was owned by The Phoenix Cement Co. Frindsbury, then Charles Eltham of Rainham, Kent. Her dimensions were 74′6″ x 15′2″ x 5′10″. P.L.A. No. 13443, registered at Rochester. She was sold to The Rev. W.H. Benson in July 1930 for £30. No known trace.

Hellvelyn Official No. 87228
Her registered tonnage was 49 reg. tons and she carried 125 tons of cement. She was built at Sittingbourne by John Masters in 1884 and was launched in April that year. She was owned by H.A. Brooks of Grays then the A.P.C.M. Her dimensions were 82′0″ x 18′9″ x 6′1″. P.L.A. No. 13403, registered at Rochester. She was sold to Mr L.H. Whiley in July 1935 for £50. A Barge Yacht in 1936. No known trace.

Henry Tinnoth Official No. 55141
Her registered tonnage was 50 reg. tons and she carried 125 tons of cement. She was built at Halling by Thomas Weekes in 1866 and launched in April of that year. She was owned by Thomas Weekes of Burham then the A.P.C.M. Her dimensions were 81′0″ x 18′6″ x 6′9″. P.L.A. No. 14443, registered at Rochester. She was sold to Chas. White of Gravesend in July 1927 for £250. Hulked in Damhead Creek 1940.

Herbert & Harold Official No. 98793
Her registered tonnage was 42 reg. tons and she carried 105 tons of cement. She was built at Rochester by William Higham in 1890 and launched in December of that year. She was owned by Henry Peters of Burham & Wouldham then H.A. Brooks of Grays, then A.P.C.M. Her dimensions were 77′0″ x 16′6″ x 5′0½″. P.L.A. No. 13415, registered in Rochester. She was sold to W.H. Theobald in December 1928 for £250. She was sunk in collision near the Chapham Head. She was raised and broken up at Rosherville circa 1930. No known trace.

Heron Official No. 110174
Her registered tonnage was 64 reg. tons and she carried 160 tons of cement. She was built at Greenhithe by F.T. Everard in 1899 and was owned by Thomas Bevan of Northfleet then the A.P.C.M. Her dimensions were 83′6″ x 20′1″ x 6′3″ (she was a big barge). P.L.A. No.

9905, registered in London. She was sold to R. Butcher in 1927 for £745. A Barge Yacht in 1947.

Hibernia Official No. 110967
Her registered tonnage was 45 reg. tons and she carried 115 tons of cement. She was built at Strood by George H. Curel in 1899 and owned by The West Kent Portland Cement Co. Burham, then the A.P.C.M. Her dimensions were 78′6″ x 18′0″ x 6′3″. P.L.A. No. 14473, registered in Rochester. She was sold to Thomas Watson, Chatham in May 1928 for £242.15s. Was a house boat in 1949.

Hilda Official No. 87054
Her registered tonnage was 42 reg. tons and she carried 80 tons of cement. She was built at Henley by Alfred Keep in 1881 and was owned by John Bazeley White of Swanscombe, then F.T. Everard in 1901. She was registered in London. No known trace.

Holborough Official No. 81893
Her registered tonnage was 41 reg. tons and she carried 80 tons of cement. She was built at Rochester by J. Gill & Son in 1880 and launched in the October of that year. She was owned by Samuel Smith of Halling, William Lee of Halling, then the B.P.C.M. and A.P.C.M. Her dimensions were 73′6″ x 14′10″ x 5′5″. P.L.A. No. 14474, registered at Rochester. She was sold to E. Watkins in September 1932 for £9. 1937 was used as a swimming raft. 1957 . Derelict in Whitewall Creek.

Hope Official No. 76594
Her registered tonnage was 44 reg. tons and she carried 90 tons of cement. She was built at Sittingbourne by R.M. Shrubsall in 1877 and was owned by John K. Anderson of Halling, then the A.P.C.M. Her dimensions were 75′8″ x 18′1″ x 5′10″. P.L.A. No. 13413, registered at Rochester. Her hulk was sold to H. Hammond in January 1933 for £3. No known trace.

Humber Official No. 108512
Her registered tonnage was 46 reg. tons and she carried a 100 tons of cement. She was built at Sittingbourne by Alf White in 1897 and owned by P. Hilton, then A.P.C.M. Her dimensions were 79′6″ x 18′0″ x 6′0″. P.L.A. No. 13416, registered in Rochester. She was sold to The Meux Breweries Co. in February 1928 for £362. Traded from Nine Elms to Rochester with beer. Finished sailing circa 1943 – 1944 was a hulk.

Invicta Official No. 58455
Her registered tonnage was 40 tons and she carried 80 tons of cement. She was built at Halling by William Lee in 1868 and owned by Samuel Smith off Halling then William Lee of Blackfriars, then the A.P.C.M. She was registered in Rochester. She was sold to Arthur Larraman of Rochester in 1923 for £260. Foundered 2-1-1925 off Chapman Head. Later raised.

Ex-W. Morgan's *Irex* circa 1920

Irex Official No. 94577
Her registered tonnage was 44 reg. tons and she carried 100 tons of cement. She was built at Rochester by J. Gill & Sons in 1889 and was launched in October of that year. She was owned by William Morgan of Halling then the A.P.C.M. Her dimensions were 76′4″ x 17′7″ x 5′9″. P.L.A. No. 13455, registered in Rochester. She was sold to E.L. Pryor of Blackheath for £100.

Ironsides Official No. 112710
Her registered tonnage is 78 tons and she carried 160 tons of cement. She was built at Grays by Clark & Standfield in 1900 and owned by the West Thurrock Lighterage Co., then the A.P.C.M. Her dimensions are 84′0″ x 20′3″ x 6′4″ and built steel. P.L.A. No. 9980, registered in London. She was sold to The London & Rochester Trading Co. in October 1927 for £550 and converted to a motor barge circa 1937. She is now a yacht barge (fully restored). NOTE: on 6-10-1929 was abandoned off Dungeness. Crew landed by lifeboat but reboarded later.

James Official No. 45519
Her registered tonnage was 42 reg. tons and she carried 90 tons of cement. She was built at Frindsbury in 1863 and owned by the Burham Brick, Cement and Lime Co. of Burham, then the A.P.C.M. Her dimensions were 72′4″ x 14′4″ x 5′2″. She was registered in Rochester. She was condemned. No known trace.

January Official No. 58477
Her registered tonnage was 36 reg. tons and she carried 80 tons of cement. She was built at Sittingbourne in 1868 and owned by the Burham Brick, Lime & Cement Co. Her dimensions were 72′0″ x 14′5″ x 4′7″. She was registered at Rochester. She was sold to Mr White of Gravesend in 1920 for £10. A canal or cut barge.

Jay Official No. 10190
Her registered tonnage was 67 reg. tons and she carried 125 tons of cement. She was built at Northfleet by Thomas Bevan in 1892 and was owned by this company then A.P.C.M. P.L.A. No. 9754, registered in London. She was sold as a wreck to Surridge in January 1927 for £50 was repaired and resold to T.F. Woods (Gravesend) circa 1935. Sailing in 1939.

J.B.W. Official No. 123813
Her registered tonnage was 72 reg tons and she carried 150 tons of cement. She was built at Millwall by Edwards & Co. in 1907 and owned by John Bazley White & Co., of Swanscombe then the A.P.C.M. P.L.A. No. 13794 registered in London. She was sold to J.W. Price in March 1927 for £850. Resold to Metcalf 1937. She was built of steel and sunk by mine 15.7.1943. Crew lost. Cargo of brick rubble.

J.M.W. Official No. 84417
Her registered tonnage was 44 reg. tons and she carried 105 tons of cement. She was built at Blackwall by A. White in 1881 and owned by Robins & Co., London Lighterage Co. then the A.P.C.M. Dimensions were 79′0″ x 16′6″ x 5′5″. P.L.A. No. 10452, registered in London. She was sold to Mr Harrison in March 1932 for £25. Derelict in Conyer Creek 1937.

John Official No. 44596
Her registered tonnage was 49 reg. tons and she carried 90 tons of cement. She was built at Strood in 1862 and owned by Edward W. Brooks of Grays, then the A.P.C.M. Her dimensions were 73′9″ x 15′7″ x 6′0″. She was registered in Rochester. She was destroyed and lost without trace.

John Official No. 47951
Her registered tonnage was 38 reg. tons and she carried 80 tons of cement. She was built at Rochester in 1864 and owned by Henry Peters of Burham & Wouldham, then the A.P.C.M. She was registered in Rochester. She was condemned and sold to Frederick Perrin of Aylesford. No known trace.

John & Edward Official No. 52926
Her registered tonnage was 44 reg. tons and she carried 100 tons of cement. She was built at Sittingbourne by Steven Taylor in 1865 and owned by The West Kent Portland Cement Co. then the A.P.C.M. Her dimensions were 75′8″ x 18′10″ x 6′7″. P.L.A. No. 13724, registered in London. Rebuilt in 1905. She was sold to F.H. Wells & Co. in May 1930 for £35. Broken up at Cross Ness.

John Little Official No. 58459
Her registered tonnage was 42 reg. tons and she carried 80 tons of cement. She was built at Upnor by W.B. Little in 1868 and owned by A.P.C.M. Her dimensions were 71′8″ x 14′8″ x 5′1″. She was registered in Rochester. She was sold to Mr Hodges in 1923 for £5. No known trace.

John Pain Official No. 84422
Her registered tonnage was 61 reg. tons and she carried 120 tons of cement. She was built at Brentford by F. White in 1881 and launched in December of that year. She was owned by Jos. Westwood, then the Tolhurst Cement Co., then the A.P.C.M. She was registered in Rochester. Sold to S.J. Brice & Son in June 1927 for £10. No known trace.

John Tinnoth Official No. 4575
Her registered tonnage was 43 reg. tons and she carried 75 tons of cement. She was built at Frindsbury in 1851 and owned by Thomas Weekes, Trenchman Weekes, then the A.P.C.M. Her dimensions were 72′1″ x 12′6″ x 4′8″. She was registered at Rochester. Sold to Mr Gobblett in 1915 for £60. An early canal or cut barge. Lime Cargo overheated 7-7-1883. Barge badly burned at Wapping.

Joseph Official No. 84407
Her registered tonnage was 41 reg. tons and she carried 80 tons of cement. She was built at Rochester by William Highams in 1881 and launched in September of that year. She was owned by Henry Peter of Burham & Wouldham. P.L.A. No. 14476, registered in Rochester. She was sunk and condemned then sold to S.A. Groom & Sons in October 1926 for £6. She was stumpie rigged. No known trace.

Judy Official No. 76601
Her registered tonnage was 42 reg. tons and she carried 80 tons of cement. She was built at Rochester in 1877 and owned by William Lee of Halling then B.P.C.M. then A.P.C.M. Her dimensions were 72′9″ x 14′7″ x 4′8″. P.L.A. No. 14477, registered in Rochester. She was sold to Mr E.J. Dowsett in June 1933 for £25. 1935 was trading to Lower Halstow with manure. No known trace.

Julia Official No. 49833
Her registered tonnage was 40 reg. tons and she carried 120 tons of cement. She was built at Rochester in 1864 and owned by George Anderson, 1885 owned by Jos. Stone of Upnor was coasting. Then Hilton Anderson & Brooks in 1901–1902 A.P.C.M. Her dimensions were 78′6″ x 18′6″ x 5′5″. She was registered at Rochester. She was condemned and sold to Joseph Stone of Upnor in February 1901.

July Official No. 58502
Her registered tonnage was 37 reg. tons and she carried 80 tons of cement. She was built at Boston, Lincs., by John Richardson in 1869

and was launched in September of that year. She was owned by Burham Brick, Lime & Cement Co., then the A.P.C.M. Her dimensions were 68′4″ x 16′0″ x 5′8″. P.L.A. No. 10507, registered at Rochester. Laid up in 1929. She was sold to Mr C. Bew in May 1930 for £7.10s. No known trace.

June Official No. 58497
Her registered tonnage was 35 reg. tons and she carried 75 tons of cement. She was built at Murston in 1869 by Steven Taylor and was owned by The Burham Brick, Lime & Cement Co., then the A.P.C.M. Her dimensions were 73′0″ x 15′1½″ x 5′5½″. P.L.A. No. 10506 registered in Rochester. She was sold to Mr Arthur Bennett in July 1933 for £50. She now lies hulked at Conyer, Kent. (Feature in Mr Arthur Bennett's delightful book 'June of Rochester').

Kate Official No. 76583
Her registered tonnage was 47 reg. tons and she carried 100 tons of cement. She was built at Rochester by Edmund G. Watson in 1876 and was owned by Robins & Co. of Northfleet then the A.P.C.M. Her dimensions were 77′0″ x 17′6″ x 5′10″. P.L.A. No. 10453, registered at Rochester. She was sold to The Canvey Supply Co. in February 1929 for £200. 1930 was a lighter. No known trace.

Kate Powell Official No. 2007
Her registered tonnage was 51 reg. tons and she carried 110 tons of cement. She was built at Maidstone in 1857 and owned by Thomas Weekes then Trenchman Weekes of Burham. Her dimensions were 68′8″ x 14′4″ x 6′3″. She was registered at Rochester, was coasting in earlier days. She was sold to The Fielder Heckman Co. in 1916 for £295. Listed as Boomsail rigged, later Topsail. 1925 was owned by BEW (Leigh).

Kestrel Official No. 89503
Her registered tonnage was 63 reg. tons and she carried 130 tons of cement. She was built at Northfleet by Thomas Bevan in 1883 and was owned by this company, then A.P.C.M. Her dimensions were 79′7″ x 18′5″ x 5′6″. She sank in collision in the Thames and was condemned and sold to Mr Mann in 1926 for £3.10s. (Cut down by a steamer in the Thames and became a total wreck).

Lady Day Official No. 113692
Her registered tonnage was 46 reg. tons and she carried 80 tons of cement. She was built by William Highams of Rochester in 1900 and was absorbed into the A.P.C.M. Was sold to J.B. Bradley in 1933 and later was used as a lighter. No known trace.

Lark Official No. 112735
Her registered tonnage was 54 reg. tons and she carried 130 tons of cement. She was built at Greenhithe by F.T. Everard in 1900 and

owned by the A.P.C.M. Her dimensions were 81′0″ x 18′4″ x 6′6″. P.L.A. No. 9896, registered in London. She was sold to J.R. Piper for £612.10s. in July 1927. She was abandoned at Dunkirk on 1st June 1940.

Larkfield Official No. 97725
Her registered tonnage was 43 reg. tons and she carried 80 tons of cement. She was built at Rochester by J. Gill & Son in 1890 and was launched in the August of that year. She was owned by Samuel Smith of Halling, William Lee of Halling then the B.P.C.M. and A.P.C.M. P.L.A. No. 14478, was registered at Rochester. She was sunk in collision on 13th December 1929. Her hulk was raised and sold to Mr Marshall in 1931 for £1. and was taken to Upnor and broken up.

Laura Official No. 67093
Her registered tonnage was 45 reg. tons and she carried 80 tons of cement. She was built at Sittingbourne by R.M. Shrubsall in 1874 and owned by John Bazley White of Swanscombe, then A.P.C.M. Her dimensions were 77′0″ x 17′6″ x 5′8″. P.L.A. No. 8988, registered in Rochester. She was sold to Mr G. Blackman in March 1925 for £5. Buried at Shorts factory Rochester circa 1932.

Ex-Francis & Co. *Lizard* having work done to her at the Upnor barge yard circa 1920

Lizard Official No. 99023
Her registered tonnage was 50 reg. tons and she carried 115 tons of cement. She was built at Rochester by Gill & Son in 1891 and owned

Ex-Francis & Co. *Lizard* hulked at Harty Ferry, Swale, 1963.

by Franics & Co. of Cliffe, Kent then the A.P.C.M. Her dimensions were 81′6″ x 18′9″ x 5′10″. P.L.A. No. 11111, registered in London.. She was sold to Thomas Watson of Rochester in February 1928 for £360 then sold to W. Surridge in 1933. She was requisitioned by H.M. Government on 22.7.1940. Her hulk lies at Sayes Court, Harty, on the Isle of Sheppey.

Loo Official No. 87051
Her registered tonnage was 52 reg. tons and she carried 90 tons of cement. She was built at Greenhithe by Alfred Keep in 1876 and was launched in August of that year. She was owned by John Bazley White of Swanscombe, then the A.P.C.M. Swimheaded. Her dimensions were 70′0″ x 18′6″ x 6′6½″. P.L.A. No. 8990, registered in London. She was sold in July 1926 to Herman Simmonds of Chatham for £180. Laid rigged at Bennett's Wharf, Gillingham for many years from 1946 and eventually fell to pieces.

Loualf Official No. 87121
Her registered tonnage was 64 reg. tons and she carried 130 tons of cement. She was built at Conyer by Alfred Marconi White in 1898 and owned by Francis Carey of Greenhithe then the A.P.C.M. Her dimensions were 86′0″ x 18′10″ x 6′10″. P.L.A. No. 11973, registered at Rochester. She was sold to A.G. Wood of Sittingbourne in May 1928 for 10s.0d. Derelict prior to 1936. Her hulk lies at Murston in Milton Creek.

Louisa Official No. 87121
Her registered tonnage was 44 reg. tons and she carried 100 tons of cement. She was built at Northfleet by The Lighterage Co. in 1883 and owned by this company then the A.P.C.M. Her dimensions were 78′0″ x 17′3″ x 6′1″. P.L.A. No. 10454, registered in London. She was sold to Mr C. White in December 1928 for £155. She was working in the Isle of Wight. She was built in Robins Creek from timber that was left over from building the 'W.M. Paxton' and 'R.A. Gibbons'. Her hulk lies at Coles Wharf, Cowes, Isle of Wight.

Lucy Official No. 52678
Her registered tonnage was 43 reg. tons and she carried 70 tons of cement. She was built at Lambeth in 1865 and was owned by John Bazley White Co. of Swanscombe then the A.P.C.M. In 1902 F.T. Eberhardt is shown as owner. Her dimensions were 67′0″ x 17′2″ x 5′1″. She was registered in London. No known trace.

Mahatma Official No. 99008
Her registered tonnage was 52 reg. tons and she carried 135 tons of cement. She was built at Bow Creek by William Lake in 1891 and owned by W. Lake of Poplar, then Hilton Anderson, then the A.P.C.M. Her dimensions were 83′3″ x 18′4″ x 6′9″. P.L.A. No. 13405 Registered in London. She was sold to Samuel Bew in July 1929, then owned by The Thames Transport Co. as a lighter. She stranded at Leigh 1932. Total loss. Her hulk lies in Smallgains Creek on Canvey Island.

March Official No. 58485
Her registered tonnage was 38 reg. tons and she carried 80 tons of cement. She was built at Rochester in 1869 and owned by The Burham Brick Lime & Cement Co., then A.P.C.M. Her dimensions were 72′4″ x 14′5″ x 5′1″. She was registered in Rochester. Was sold to a Mr. Brown in December 1926 for £20. A canal or cut barge. No known trace.

Marconi Official No. 114679
Her registered tonnage was 43 reg. tons and she carried 115 tons of cement. She was built at Conyer by Alfred Marconi White in 1901 and was owned by J.G. Hammond of London, then R. West and came into A.P.C.M. ownership in 1922. Her dimensions were 85′6″ x 18′6″ x 5′8″. P.L.A. No. 12947, registered in London. She was sold to C.E. Wilson for £145 in December 1934. Was a yacht in Ramsgate Harbour, sank and was broken up in 1946.

Marie Official No. 76591
Her registered tonnage was 43 reg. tons and she carried 80 tons of cement. She was built at Rochester by F. Sollett in 1877 and was launched in the January of that year. She was owned by John P. Wood of Murston, Henry Peters of Wouldham & Burham, then the A.P.C.M. Her dimensions were 72′5″ x 14′6″ x 4′7″. P.L.A. No. 14480 registered in Rochester. She was sold to The Union Lighterage Co. in 1923 for £75. A canal or cut barge. No known trace.

Mary Ann Official No. 47253
Her registered tonnage was 22 reg. tons and she carried 50 tons of cement. She was built at Landport in 18? and registered at Cowes, Isle of Wight. No known trace.

Mary Fox Official No. 9404
Her registered tonnage was 37 reg. tons and she carried 70 tons of

cement. She was built at Rochester by J. Gill & Son in 1853 and was owned by Frederick Sollit of Rochester, then the A.P.C.M. Her dimensions were 71′1″ x 14′4″ x 4′6″. She was registered at Rochester. She was condemned and broken up. A canal or cut barge. No known trace.

Maud Official No. 67099
Her registered tonnage was 42 reg. tons and she carried 85 tons of cement. She was built at Milton in 1875 by R.M. Shrubsall of Milton and owned by Samuel B. Booth of Borstal then the A.P.C.M. No known trace.

Medina Official No. 120413
Her registered tonnage was 42 reg. tons and she carried 80 tons of cement. She was built at Portsmouth by Dyble in 1905 and owned by the A.P.C.M. Her dimensions were 70′0″ x 17′0″ x 4′6″. She was registered at Southampton. She was sold to The Williams Shipping Co. (Fowley Ltd) in May 1937 for £500. No known trace.

Medina Official No. 84412
Her registered tonnage was 42 reg. tons and she carried 80 tons of cement. She was built at East Cowes by J.S. White in 1881 and launched in the October of that year. Her dimensions were 74′6″ x 14′8″ x 5′6″ (composite construction). P.L.A. No. 14481, registered at Rochester. She was owned by Samuel Smith of Halling, William Lee of Halling then the A.P.C.M. She was sold to Mr Mann in 1926 for £5. No known trace.

Medway Official No. 50308
Her registered tonnage was 39 reg. tons and she carried 80 tons of cement. She was built at Faversham by J.M. Goldfinch in 1864 and owned by Geo. K. Anderson of Frindsbury, then the A.P.C.M. Her dimensions were 72′0″ x 16′6″ x 5′7″. P.L.A. No. 13422, registered in Rochester. She was sold to Wadhams & Smeed in June 1928 for £12.10s. as a lighter.

Medway Official No. 84435
Her registered tonnage was 45 reg. tons and she carried 85 tons of cement. She was built at East Cowes by J.S. White in 1882 and owned by Samuel Smith of Halling, William Lee of Halling, then the A.P.C.M. Her dimensions were 72′6″ x 14′10″ x 5′3″. P.L.A. No. 14482, registered in Rochester. She was sold to E. Baker in August 1926 for £290. Composite construction. No known trace.

Mendip Official No. 98102
Her registered tonnage was 82 reg. tons and she carried 165 tons of cement. She was built at Bow Creek by W. Lake in 1890 and owned by E.W. Brooks of Grays, then the A.P.C.M. She was registered in London. Could have been a converted lighter. No known trace.

Merlin Official No. 89635
Her registered tonnage was 62 reg. tons and she carried 120 tons of cement. She was built at Northfleet by Thomas Bevan in 1884, was owned by this company, then A.P.C.M. Her dimensions were 80′0″ x 19′0″ x 6′3″. P.L.A. No. 9751, registered in London. She was sold to J.S. Rayfield, Northfleet in January 1928 for £427. Registry closed in 1949 as broken up.

Mersey Official No. 89866
Her registered tonnage was 44 reg. tons and she carried 100 tons of cement. She was built at Faversham by J.M. Goldfinch in 1891 and launched in the April of that year. She was owned by Hilton Anderson of Upnor, then the A.P.C.M. Her dimensions were 78′6″ x 17′9″ x 5′10″. P.L.A. No. 13409, registered in Faversham. She was sold to D. Campbell in August for 1930 for £200. and converted to a barge yacht. Registry closed in 1953 as 'sunk'.

Mersey Official No. 84406
Her registered tonnage was 42 reg. tons and she carried 80 tons of cement. She was built at Rochester by Gill & Son in 1881 and launched in September of that year. She was owned by Henry Peters of Wouldham then the B.P.C.M. Her dimensions were 74′2″ x 14′9″ x 5′0″. She was stumpy rigged. P.L.A. No. 14483, registered at Rochester. She was sold to Mr P.T. Baker of The Medway Catchment Board for £75. Derelict south side Elmly Ferry 1946, burnt in 1950.

Meteor Official No. 58490
Her registered tonnage was 36 reg. tons and she carried 80 tons of cement. She was built at Murston by the B.B.C. & L. Co., in 1869. She was owned by The Burham Brick, Cement & Lime Co., then the A.P.C.M. She was registered in Rochester. No known trace.

Midsummer Official No. 113695
Her registered tonnage was 46 reg. tons and she carried 110 tons of cement. She was built by William Highams of Rochester in 1901 and registered in Rochester. Came into the ownership of the A.P.C.M. Sold in 1930s to J.B. Bradley of London. Was used as a lighter. Reported broken up at Crossness 1952.

Minicoy Official No. 96690
Her registered tonnage was 50 reg. tons and she carried 115 tons of cement. She was built at Rochester by J. Gill & Son in 1890 and owned by Francis & Co. of Vauxhall then the A.P.C.M. Her dimensions were 80′6″ x 18′6″ x 5′9″. P.L.A. No. 11110, registered in London. She was sold to Thomas Watson of Chatham in February 1928 for £360 then she went to Surridge in 1933. No known trace.

Monday Official No. 84443
Her registered tonnage was 45 reg. tons and she carried 100 tons of

cement. She was built by The Burham Brick, Lime & Cement Co. at Burham in 1882, she was owned by this company, then the A.P.C.M. Her dimensions were 79′0″ x 17′7½″ x 5′11″. P.L.A. No. 10517, registered in Rochester. Her hull only was sold to P. Harrison in March 1932 for £25. Derelict at Queenborough 1939.

Monkwood Official No. 109930
Her registered tonnage was 46 reg. tons and she carried 90 tons of cement. She was built at Rochester by Geo. Curel in 1899 and owned by Henry Peters of Wouldham, then the B.P.C.M. Her dimensions were 74′6″ x 15′0″ x 5′8½″. P.L.A. No. 14484, registered at Rochester. She was sold to A.G. Wood in September 1936 for £1. She was stumpie rigged. Was a roads barge at Greenwich from September 1929. Derelict in 1936 near Adelaide Dock.

Moorhen Official No. 89555
Her registered tonnage was 63 reg. tons and she carried 120 tons of cement. She was built at Swanscombe by John Bazley White in 1884 and was owned by John Bazley White of Swanscombe then the A.P.C.M. Her dimensions were 84′0″ x 19′8″ x 6′0″. P.L.A. No. 8976, registered in London. She was sold to William Cory & Son Ltd in October 1930 for £170 then G.A. Minnerly of Maidenhead. Registry closed in 1951 as 'broken up'.

Murston Official No. 58436
Her registered tonnage was 35 reg. tons and she carried 80 tons of cement. She was built at Sittingbourne by George Smeed in 1868 and was owned by The Burham Brick, Lime & Cement Co. Her dimensions were 71′2″ x 14′5″ x 4′5″. She was registered at Rochester. She was sold to The Union Lighterage Co. in 1923 for £80. A canal or cut barge. No known trace.

Nan Official No. 105890
Her registered tonnage was 47 reg. tons and she carried 90 tons of cement. She was built at Swanscombe by John Bazley White in 1896 and was owned by John Bazley White of Swanscombe and Gillingham, then A.P.C.M. Her dimensions were 80′6″ x 18′6″ x 5′1½″. P.L.A. No. 9115, registered in London. She was sold to the Dirt Co. in November 1931 for £125. She was sunk off Port Victoria in 1938. This barge was built for the clay work into the Gillingham Cement Works. She was known as a 'Flat-iron' build and could carry 100 tons of clay in five foot of water. No known trace.

Nashenden Official No. 108517
Her registered tonnage was 44 reg. tons and she carried 85 tons of cement. She was built at Rochester in 1898 by The Acorn Barge Co. and owned by Booth & Co. of Borstal, then the A.P.C.M. Her dimensions were 75′0″ x 14′11″ x 5′7″. P.L.A. No. 13450 registered in Rochester. She was sold to A.G. Wood of Sittingbourne in September

1933 for £2. Was a roads barge at Greenwich from June 1930. Burnt at Adelaide Dock, Murston 1949.

Needles Official No. 99043
Her registered tonnage was 49 reg. tons and she carried 115 tons of cement. She was built at Rochester by J. Gill & Son in 1892 and launched in July of that year. She was owned by Francis & Co. of Vauxhall then the A.P.C.M. Her dimensions were 80′6″ x 18′6″ x 5′10″. P.L.A. No. 11112, registered in London. She was sold to Lambert Bros. & Co. of Southampton in October 1931 for £150. Register closed in 1949 – broken up.

New Ada Official No.
Her registered tonnage was 48 reg. tons and she carried 90 tons of cement. She was built at Sittingbourne by R.M. Shrubsall in 1877 and launched in the April of that year. She was owned by John Bazley White of Swanscombe, the A.P.C.M. Her dimensions were 81′6″ x 14′8″ x 5′2″. P.L.A. No. 8986, registered at Rochester. She was sold to S. Waghorn of Chatham in May 1934 for £2. Hulked at the Upnor Barge Yard 1930.

Ex-John Bazley White's *New World* hulked at Sayes Court Harty 1960

New World Official No. 76624
Her registered tonnage was 50 reg. tons and she carried 125 tons of cement. She was built at Sittingbourne by R.M. Shrubsall in 1877 and owned by John Bazley White of Swanscombe. She was rebuilt in 1906 at Upnor and owned by the A.P.C.M. Her dimensions were 78′0″ x 19′2″ x 6′10″. P.L.A. No. 8982, registered in Rochester. She was sold to William Cory & Son Ltd. in October 1930 for £170. Still sailing 1938. Register closed 1947.

Ninety Official No. 97730
Her registered tonnage was 43 reg. tons and she carried 75 tons of cement. She was built at Rochester by Gill & Sons in 1890 and was launched in September of that year. She was owned by Henry Peters of Wouldham, Peters Cement Co. Burham, then A.P.C.M. Her dimensions were 74′0″ x 14′8″ x 5′4″. She was stumpie rigged. She was registered at Rochester. Sold to Smeed & Wadhams in March 1926 for £5. Derelict 1937–47 at Frindsbury.

Ninety Nine Official No. 110979
Her registered tonnage was 57 reg. tons and she carried 140 tons of cement. She was built at Frindsbury by Geo. H. Curel in 1900 and owned by The Peters Cement Co., then B.P.C.M., A.P.C.M. P.L.A. No. 14486, registered at Rochester. She was sold to A.J. Peters of Southend in November 1927 for £630. In 1944 Whiting Bros. sold her to Charles Burley of Sittingbourne. Had a motor fitted. Her hulk lies at Churchfields, Milton Creek, Sittingbourne.

November Official No. 76593
Her registered tonnage was 45 reg. tons and she carried 85 tons of cement. She was built at Aylesford by The Burham Brick Lime & Cement Co. and launched in December. She was owned by The B.B.L.C. Co., then the A.P.C.M. Her dimensions were 77′6″ x 17′3″ x 5′3″. P.L.A. No. 10511, registered in Rochester. Sold to J.C. Higgs & Sons in March 1928 for £60 as a lighter. No known trace.

Octavious
A bit of a mystery about this barge. The A.P.C.M. did own a barge of this name but no details were recorded only that she was sold to J.T. Rayfield in 1901 for £7.15s. No known trace.

Osprey Official No. 56751
Her registered tonnage was 54 reg. tons and she carried 105 tons of cement. She was built at Milton by R.M. Shrubsall in 1868 and owned by Thomas Bevan of Northfleet, then the A.P.C.M. She was registered in London. She was condemned.

Overcome Official No. 109929
Her registered tonnage was 44 reg. tons and she carried 90 tons of cement. She was built by Gill at Rochester in 1899 and owned by Henry Peters of Burham, then B.P.C.M., A.P.C.M. Her dimensions were 74′3″ x 14′11″ x 5′7½″. Stumpie rigged. P.L.A. No. 14487 registered in Rochester. She was sold to W.D. Inglis of Gravesend in April 1933 for £60. No known trace.

Perseverance Official No. 60892
Her registered tonnage was 39 reg. tons and she carried 80 tons of cement. She was built at Northfleet in 1869 and owned by Thomas Bevan of Northfleet, then B.P.C.M., A.P.C.M. She was registered in London. Was sold to The Union Lighterage Co. in 1923 for £75. She was a clay barge at Murston Cement Works. No known trace.

Phoenix Official No. 90970
Her registered tonnage was 51 reg. tons and she carried 130 tons of cement. She was built at Frindsbury by George H. Curel in 1884 and was launched in November of that year. She was owned by The Phoenix Barge Co. of Rochester, then H.A. Cunis of Blackfriars, then the

Ex-Peters *Overcomer* circa 1933.

A.P.C.M. Her dimensions were 81′6″ x 19′0″ x 6′1″. P.L.A. No. 13442, registered in Rochester. She was sold to A.G. Wood of Sittingbourne in January 1935 for £30. She was burnt in Adelaide Dock in 1949.

Plinlimmon Official No. 91902
Her registered tonnage was 45 reg. tons and she carried 110 tons of cement. She was built at Frindsbury by George H. Curel in 1886 and launched in February that year. She was owned by E.W. Brooks of Grays, then A.P.C.M. Her dimensions were 79′0″ x 18′0″ x 6′1″. P.L.A. No. 13408, registered in London. She was sold to a Captain Pollon for £225. Became a Yacht Barge with an engine. Still on the 1947 register.

Plover Official No. 110026
Her registered tonnage was 62 reg. tons and she carried 110 tons of cement. She was built at Swanscombe by John Bazley White in 1898 and was owned by this company, then A.P.C.M. Her dimensions were 86′0″ x 19′3″ x 6′0″. P.L.A. No. 9094, registered in London. She was sold to S.J. Brice & Sons in September 1931 for £250, worked in the clay traffic. Was a very slow sailer. Hulked at White Wall Creek, Frindsbury, Kent.

Porpoise Official No. 84397
Her registered tonnage was 44 reg. tons and she carried 80 tons of cement. She was built at West Cowes by J. White & Co. in 1881 and owned by Samuel Smith of Halling, William Lee of Blackfriars, then the A.P.C.M. Her dimensions were 74′6″ x 14′9″ x 5′10″. Composite construction. P.L.A. No. 14488, registered in Rochester. She was sold to Mr G. Mann in September 1926 for £5. No known trace.

Prince Official No. 58464
Her registered tonnage was 34 reg. tons and she carried 70 tons of cement. She was built at Battersea in 1868 and owned by H.J. Spooner of Bobbing, Kent, then the A.P.C.M. She was registered in Rochester. Was sold to W.C. White in 1919 for £105. No known trace.

Progress Official No. 73664
Her registered tonnage was 41 reg. tons and she carried 80 tons of cement. She was built at Northfleet by Thomas Bevan in 1876 and was owned by this company, then A.P.C.M. Her dimensions were 72′9″ x 14′8″ x 4′9″. She was registered in London. Was sold to W. Smith in June 1930 for £3.10s. No known trace.

Providence Official No. 74819
Her registered tonnage was 38 reg. tons and she carried 80 tons of cement. She was built at Rochester by William Highams in 1876 and was launched in August of that year. She was owned by Henry Peters of Burham & Wouldham, then the A.P.C.M. She was registered in Rochester. This vessel was condemned and broken up without trace.

Punch Official No. 76586
Her registered tonnage was 40 reg. tons and she carried 80 tons of cement. She was built at Rochester by J. Gill & Son in 1876 and was launched in December of that year. She was owned by Samuel Smith of Halling, William Lee of Blackfriars, B.P.C.M., then A.P.C.M. Her dimensions were 74′6″ x 15′0″ x 5′4″. P.L.A. No. 14489, registered in Rochester. She was sold to Mr. Cooper in June 1927 for £375. No known trace.

Quail Official No. 102819
Her registered tonnage was 79 reg. tons and she carried 180 tons of cement. She was built at Northfleet by Thomas Bevan in 1894 and was owned by this company. Her dimensions were 82′0″ x 21′6″ x 6′6½″. She was registered in London. Was sunk and condemned early in her life. Was one of the largest barges in the A.P.C.M. Fleet. Sunk in collision 6.10.1911 in the Thames.

R.A. Gibbons Official No. 85161
Her registered tonnage was 55 reg. tons and she carried 145 tons of cement. She was built at Northfleet by The Lighterage Co. in 1882 and owned by Robins & Co. Northfleet, then A.P.C.M. Her dimensions were 84′0″ x 19′6″ x 6′0″. P.L.A. No. 10455, registered in London. She was sold to Thomas Watson of Chatham. 1934 to L. Surridge. Derelict at Belvedere 1940.

Rathmona Official No. 106529
Her registered tonnage was 45 reg. tons and she carried 115 tons of

cement. She was built at Sittingbourne by R.M. Shrubsall in 1897 and owned by Geo. Austin of Northfleet, Tolhurst Cement Co. then B.P.C.M., A.P.C.M. P.L.A. No. 8323, registered in Rochester. She was sold to H.G. Warpole in September 1929 for £75 (as a wreck) then passed to Rayfield of Northfleet. Originally stumpy rigged. Later topsail. Hulked 1940.

Raven Official No. 94387
Her registered tonnage was 64 reg. tons and she carried 130 tons of cement. She was built at Northfleet by T. Bevan & Co, in 1888 and was owned by this company, then A.P.C.M. Was registered in London. Was a 'swimmie'. Finished circa 1918. No known trace.

Regent Official No. 105805
Her registered tonnage was 35 reg. tons and she carried 70 tons of cement. She was built at Swanscombe by J.B. White in 1896 and was owned by John Bazley White of Swanscombe then A.P.C.M. Her dimensions were 73′0″ x 14′5″ x 4′9″ – was stumpie rigged. P.L.A. No. 9008 registered in London. She was sold to H.I. Thomas in June 1932 for £40 – this was hull only. Was a mooring vessel at Silvertown 1940. No known trace.

Renown Official No. 55183
Her registered tonnage was 38 reg. tons and she carried 70 tons of cement. She was built by William Lee at Halling in 1867 and was owned by Samuel Smith of Halling, William Lee of Blackfriars, then A.P.C.M. Her dimensions were 72′7″ x 14′5″ x 4′9″. P.L.A. No. 14454, registered in Rochester. Sold to J.T. Stratford in January 1926 for £20. A canal or cut barge. No known trace.

Resurga Official No. 118464
Her registered tonnage was 50 reg. tons and she carried 130 tons of cement. She was built at Conyer by A.M. White in 1904 and owned by Samuel West of Gravesend, The Kent Portland Cement Co., then A.P.C.M. Her dimensions were 84′1″ x 19′0″ x 6′0″. P.L.A. No. 13723, registered in London. She was sold to William Cory & Sons Ltd. in October 1930 for £170. She had a 'half moon' or Cruiser stern. Dummy barge at Charlton 1949.

Rover Official No. 52938
Her registered tonnage was 38 reg. tons and she carried 80 tons of cement. She was built at Halling by W. Lee in 1865 and was owned by Samuel Smith of Halling, William Lee of Blackfriars, then A.P.C.M. Her dimensions were 71′4″ x 14′6″ x 5′0″ and had round bows. She was registered in Rochester. Was sold to G.G. Francis in 1921 for £60. No known trace.

Sapphire Official No. 84226
Her registered tonnage was 51 reg. tons and she carried 110 tons of

cement. She was built at Landport, Isle of Wight in 1880 and owned by John Bazley White & Co. then the A.P.C.M. She was registered in Faversham (1884). 1885 owner F. Wagher (Herne Bay). Could have been used as a lighter. On the 1921 Register as unrigged. No known trace.

Sarah Official No. 23425
Her registered tonnage was 43 reg. tons and she carried 75 tons of cement. Built in 1860. This vessel was condemned and broken up without trace. No other details.

Scarborough Official No. 87205
Her registered tonnage was 41 reg. tons and she carried 80 tons of cement. She was built at Rochester by William Higham in 1882 and launched in the October of that year. She was owned by Henry Peters of Burham, then B.P.C.M. Her dimensions were 74′0″ x 15′0″ x 5′5″. Was stumpy rigged. P.L.A. No. 14491 registered in Rochester. She was sold to Braithwaite & Co in March 1933 for £30. Foundered in Sea Reach 21.10.34. 1953 derelict at Wallasea Island.

Seagull Official No. 108350
Her registered tonnage was 54 reg. tons and she carried 120 tons of cement. She was built at Greenhithe by John Bazley White in 1898 and was owned by John Bazley White of Swanscombe, then A.P.C.M. P.L.A. No. 9142, registered in London. She was sold to Thomas Watson of Chatham in February 1928 for £400. Later owned by Surridge. Finished sailing in 1940. No known trace.

Seagull Official No. 56871
Her registered tonnage was 51 reg. tons and she carried 110 tons of cement. She was built at Faversham in 1867 and owned by Thomas Bevan of Northfleet, then the A.P.C.M. She was registered in London. Was sold to Robinson & Son in 1918 for £5. No known trace.

September Official No. 67089
Her registered tonnage was 46 reg. tons and she carried 100 tons of cement. Was built at Aylesford by The Burham Brick, Lime & Cement Co. in 1874 and was owned by this company, then A.P.C.M. Her dimensions were 78′3″ x 17′6″ x 5′9″. P.L.A. No. 10509 registered in Rochester. She was in collision and sunk. Was condemned, then sold to a Mr A. Pollard in June 1929 for £3. Reported broken-up Norton's Yard.

Severn Official No. 94560
Her registered tonnage was 45 reg. tons and she carried 100 tons of cement. She was built at Rochester by Gill & Son in 1888 and launched in August of that year. She was owned by Hilton Anderson of Halling, then A.P.C.M. Her dimensions were 79′6″ x 17′6″ x 5′7″. P.L.A. No. 13414, registered at Rochester. She was sold to R. Norton Jnr. in

November 1935 for £15. Reported broken up at Bugsbys Hole, East Greenwich.

Shannon Official No. 109920
Her registered tonnage was 57 reg. tons and she carried 140 tons of cement. She was built at Sittingbourne by A. White in 1898 and owned by Hilton Anderson of Halling, then the A.P.C.M. Her dimensions were 86′7″ x 20′6″ x 6′2″. P.L.A. No. 13419, registered in Rochester. She was sold to A.J. Peters of Southend in July 1928 for £530. Finished sailing in 1941. No known trace.

Shark Official No. 84383
Her registered tonnage was 44 reg. tons and she carried 80 tons of cement. She was built at East Cowes by J.S. White in 1881 and owned by Samuel Smith of Halling, William Lee of Blackfriars, then A.P.C.M. Her dimensions were 75′0″ x 15′0″ x 5′9″. P.L.A. No. 14492, registered in Rochester. She was sunk and condemned, then sold to a Mr Mann in 1926 for £7. No known trace.

Muddies loading *Silica* in Damhead Creek circa 1900

Silica Official No. 110961
Her registered tonnage was 54 reg. tons and she carried 135 tons of cement. She was built at Milton by A. White in 1899 and owned by Peters Cement Co., then A.P.C.M. Her dimensions were 78′4″ x 21′5″ x 5′0″. P.L.A. No. 14493, registered at Rochester. She was sold to S.J. Brice & Son in November 1928 for £728. Foundered 2.6.1938. Derelict on Stoke Marshes after working as a clay barge 1941.

Ex-Peter's *Silica* hulked outside Damhead Creek 1963

Silver Wedding Official No. 77032
Her registered tonnage was 53. reg. tons and she carried 90 tons of cement. She was built at Charlton by William Cory in 1877 and owned by Thomas Bevan of Northfleet, then A.P.C.M. Her dimensions were 74′6″ x 15′3″ x 5′3″. P.L.A. No. 9747 registered in London. She was sold to A.D. Hippisby Cox in June 1933 for £70. No known trace.

Silver Wedding Official No. 106526
Her registered tonnage was 50 reg. tons and she carried 135 tons of cement. She was built at Rochester by William Highams in 1896 and owned by J.M. Wakefield of Northfleet, then Tolhurst Cement Co., then A.P.C.M. Her dimensions were 80′0″ x 19′0″ x 5′6″. P.L.A. No. 11220, registered in Rochester. She was sold to York Stoneham Co. in 1927 for £600 then Mr Ian Hassell in 1933 for £155. No known trace.

Sincerity Official No. 110074
Her registered tonnage was 55 reg. tons and she carried 120 tons of cement. She was built at Rochester in 1899 and was owned by The West Thurrock Lighterage Co. and then A.P.C.M. She was registered in London. No known trace.

Sophie Official No. 56894
Her registered tonnage was 33 reg. tons and she carried 62 tons of cement. She was built at Reading in 1865 and owned by George Earle of Brentford. She was registered in London. This vessel was condemned and broken up without trace.

Spider Official No. 118265
Her registered tonnage was 72 reg. tons and she carried 185 tons of cement. She was built in 1903 and registered in London. This vessel was probably a converted lighter. She was sold to E.L. Wilson in May 1930 for £290. Reported broken up at Crossness.

Spring Official No. 79886
Her registered tonnage was 46 reg. tons and she carried 80 tons of cement. She was built at Aylesford by The Burham Brick, Lime & Cement Co. in 1879 and was owned by this company then the A.P.C.M. Her dimensions were 74′3″ x 14′6″ x 4′8½″. P.L.A. No. 10513, registered in Rochester. She was sold to Short Bros. of Rochester in January 1932 for £6. Reported buried with others at Shorts Works, Rochester.

Spry Official No. 77094
Her registered tonnage was 44 reg. tons and she carried 80 tons of cement. She was built at Limehouse by Surridge & Hartnell in 1878 and owned by Thomas Bevan of Northfleet, then the A.P.C.M. She was registered in London. Was sold to Mr Hodges in 1923 for £5. No known trace.

Stanley Margetts Official No. 79878
Her registered tonnage was 44 reg. tons and she carried 90 tons of cement. She was built at Frindsbury by Geo. Curel in 1878 and launched in the December of that year. She was owned by The West Kent Cement Co. Burham, then the A.P.C.M. Her dimensions were 79′0″ x 17′3″ x 5′9″. P.L.A. No. 14494 registered in Rochester. She was badly damaged in a collision with a L.C.C. Hopper on 29.1.1931. Was sold to C. Bews & Son in May 1931 for £30. No known trace.

Stratford Official No. 81899
Her registered tonnage was 42 reg. tons and she carried 80 tons of cement. She was built at Frindsbury by Geo. Curel in 1880 and owned by Thomas Weekes of Halling, then Trenchman Weekes of Halling, then the A.P.C.M. Her dimensions were 76′4″ x 15′0″ x 4′1″. She was registered in Rochester. She was derelict at Northfleet. No known trace.

Success Official No. 68494
Her registered tonnage was 40 reg. tons and she carried 80 tons of cement. She was built at Northfleet by Thomas Bevan and was owned by this company then the A.P.C.M. Her dimensions were 73′0″ x 14′4″ x 5′0″. She was registered in London. A canal or cut barge. No known trace.

Summer Official No. 81862
Her registered tonnage was 42 reg. tons and she carried 80 tons of cement. She was built at Aylesford by The Burham Brick, Lime & Cement Co., who also owned her, then the A.P.C.M. Her dimensions were 75′11″ x 15′2″ x 5′9″. P.L.A. No. 10514, registered in Rochester –sank in collision 16.11.1927. She was sold to Mr Moss in February 1928 for £35. No known tracc.

Surrey Official No. 58457
Her registered tonnage was 37 reg. tons and she carried 80 tons of

cement. She was built at Rochester by M. Gill & Son in 1868 and owned by the Burham Brick Lime & Cement Co., then A.P.C.M. Her dimensions were 72′5″ x 14′5″ x 4′6″. Registered in Rochester. She was sold to The Union Lighterage Co. in 1923 for £80. A canal or cut barge. No known trace.

Sussex Official No. 55193
Her registered tonnage was 42 reg. tons and she carried 90 tons of cement. She was built at Frindsbury by Geo. Curel in 1867 and owned by The Burham Brick, Lime & Cement Co., then the A.P.C.M. She was registered in Rochester 1919–25 owned by H.A. Cunis. This vessel was condemned. No known trace.

Swale Official No. 104932
Her registered tonnage was 48 reg. tons and she carried 110 tons of cement. She was built at Faversham by Goldfinch in 1894 and owned by Hilton Anderson, then A.P.C.M. Her dimensions were 78′0″ x 18′0″ x 5′11″. P.L.A. No. 13424 registered in Faversham. She was sold to Mr Hollingsberry in October 1933 for £150. Yacht barge 1938. No known trace.

The Swan Official No. 87050
Her registered tonnage was 59 reg. tons and she carried 120 tons of cement. She was built at Swanscombe by J.B. White in 1882 and owned by John Bazley White, then the A.P.C.M. She was registered in London. No known trace.

Swift Official No. 65622
Her registered tonnage was 41 reg. tons and she carried 80 tons of cement. She was built at Northfleet by Francis Barrett in 1871 and launched in August that year. She was owned by Thomas Bevan of Northfleet then the A.P.C.M. Her dimensions were 73′8″ x 15′4″ x 5′9″. P.L.A. No. 9738, registered in London. She was sold to The Wandsworth Gas Co. in March 1929 for £150. (She was stumpie rigged). No known trace.

Tay Official No. 109915
Her registered tonnage was 46 reg. tons and she carried 115 tons of cement. She was built at Sittingbourne by A. White in 1898 and owned by Hilton Anderson of Halling, then the A.P.C.M. Her dimensions were 79′0″ x 17′10″ x 6′0″. P.L.A. No. 13418, registered in Rochester. She was sold to The Supply & Transport Co. Ltd. in October 1930 for £225. Later owned by Surridge. No known trace.

Teal Official No. 91890
Her registered tonnage was 53 reg. tons and she carried 120 tons of cement. She was built at Swanscombe by John Bazley White in 1885 and was owned by this company then the A.P.C.M. Her dimensions were 83′3″ x 18′9″ x 6′5″. P.L.A. No. 8979, registered in London.

She was sold to C. Bew & Son in February 1931 for £180 then went to L. Surridge.

Thursday Official No. 94557
Her registered tonnage was 44 reg. tons and she carried 90 tons of cement. She was built at Aylesford by The Burham, Brick Lime & Cement Co. in 1881 and was launched in June of that year. She was owned by this company, then the A.P.C.M. Her dimensions were 77′9″ x 16′9″ x 5′10″. P.L.A. No. 10520, registered in Rochester. She was sold to Mr J. Hone in August 1934 for £90. Yacht barge 1934. Reported foundered off Anglesey.

Tinzenhorn Official No. 96696
Her registered Tonnage was 50 reg. tons and she carried 130 tons of cement. She was built at Blackwall by W. Lake in 1890 and owned by A.E. Brooks of Grays, then A.P.C.M. Her dimensions were 84′0″ x 19′6″ x 6′8″. P.L.A. No. 13406, registered in London. She was sold to Mr. Jones in March 1933 for £40. Hulk at Greenwich Power Station 1939.

Titbits Official No. 99920
Her registered tonnage was 49 reg. tons and she carried 100 tons of cement. She was built at Rochester in 1892 and was owned by G.T. Earle & Co, then the A.P.C.M. She was registered in Rochester. Was sunk and condemned then sold to S.A. Erin in October 1926 for £6. No known trace.

The hulk of the B.B.C. & L. Co. *Tuesday* at Orford Essex 1953. She is the only remaining hulk of the barges that were named after the days of the week, months of the year, in recognisable form.

Tuesday Official No. 87222
Her registered tonnage was 42 reg. tons and she carried 85 tons of cement. She was built at Aylesford by The Burham Brick, Lime & Cement Co., in 1883 and launched in November of that year. She was owned by this company, then A.P.C.M. Her dimensions were 73′6″ x 15′2″ x 6′0″. P.L.A. No. 10518, registered in Rochester. Sold to a Captain Skinner in May 1930 for £100. Hulked at Orford, Essex.

Two Brothers Official No. 27619
Her registered tonnage was 35 reg. tons and she carried 70 tons of cement. She was built at Rochester by William Highams in 1860 and owned by Thomas Weekes, then Trenchman Weekes, then the A.P.C.M. Her dimensions were 72′7″ x 13′5″ x 4′9″. She was registered in Rochester. Sold to Mann Baker in December 1913 for £18. A canal barge. This barge had code Flags 'PSCG'. Was probably coasting when she was new.

Tyne Official No. 60237
Her registered tonnage was 41 reg. tons and she carried 80 tons of cement. She was built at Faversham by Goldfinch in 1869 and owned by John B. Anderson then the A.P.C.M. Her dimensions were 73′0″ x 17′3″ x 5′8″. P.L.A. No. 13401 registered in Faversham. She was sold to Talbot & Co. in August 1927 for £55. No known trace.

Tyne Official No. 87219
Her registered tonnage was 42 reg. tons and she carried 80 tons of cement. She was built at East Cowes by J.S. White in 1883 and launched in March of that year. She was owned by Samuel Smith of Halling, William Lee of Halling then B.P.C.M., A.P.C.M. Composite construction. P.L.A. No. 14496, registered in Rochester. This vessel was condemned and lost without trace.

Ural Official No. 101907
Her registered tonnage was 80 reg. tons and she carried 150 tons of cement. She was built in Canning Town, London in 1892 and owned by Hilton Anderson & Brooks, then A.P.C.M. She was registered in London. (Was a large swim headed lighter that had been rigged). Lost or broken up without trace.

Urgent Official No. 76629
Her registered tonnage was 41 reg. tons and she carried 80 tons of cement. She was built at Rochester by Gill & Son in 1877 and launched in August. She was owned by Samuel Smith of Halling then B.P.C.M. Her dimensions were 73′0″ x 15′0″ x 5′7″. P.L.A. No. 14497, registered in Rochester. She was sold to H.G. Baker in February 1933 for £60. No known trace.

Varnes Official No. 58450
Her registered tonnage was 41 reg. tons and she carried 90 tons of cement. She was built at Frindsbury by Geo. Curel in 1868 and owned by The Burham Brick Lime & Cement Co., then the A.P.C.M. Her dimensions were 74′0″ x 16′6″ x 5′10″. P.L.A. No. 10530, registered in Rochester. Was sunk in collision with S.S. Cambridge in May 1932, Halfway Reach in the River Thames.

Venus Official No. 58452
Her registered tonnage was 37 reg. tons and she carried 75 tons of

cement. She was built at Sittingbourne in 1868 and owned by The Burham Brick Lime & Cement Co., then the A.P.C.M. Her dimensions were 74′6″ x 15′0″ x 5′9″. She was registered in Rochester. Sold to Mr G. Mann in 1926 for £7. No known trace.

Victoria Official No. 108338
Her registered tonnage was 44 reg. tons and she carried 110 tons of cement. She was built at Rochester by William Highams in 1898 and owned by H. Austen of Rochester, then A.P.C.M. Her dimensions were 79′5″ x 18′6″ x 5′6″. Stumpie rigged. P.L.A. No. 9116, registered in Rochester. This barge was badly damaged in collision with the S.S. 'Dalgarth Force' in April 1931 and sold to Wrightson & Son Ltd. without her gear for £45. Reported used as a collar barge. No known trace.

Victoria Official No. 108239
Her registered tonnage was 45 reg. tons and she carried 90 tons of cement. She was built at Greenhithe by John Bazley White in 1897 and owned by John Bazley White, Swanscombe, then the A.P.C.M. Her dimensions were 79′5″ x 18′6″ x 5′6″. P.L.A. No. 9955, registered in London. She was sold to Lambert Bros. in March 1932 for £75. No known trace.

Wasp Official No. 52968
Her registered tonnage was 38 reg. tons and she carried 90 tons of cement. She was built at Sittingbourne in 1866 and owned by The Burham Brick Lime & Cement Co., then A.P.C.M. P.L.A. No. 10502, registered in Rochester. She was sold to Bav. & Co. in October 1928 for £50. No known trace.

Wednesday Official No. 90973
Her registered tonnage was 41 reg. tons and she carried 80 tons of cement. She was built at Aylesford by The Burham Brick Lime & Cement Co. in 1885 and owned by this company, then the A.P.C.M. Her dimensions were 73′0″ x 15′0″ x 5′6″. Stumpy barge. P.L.A. No. 10519, registered in Rochester. She was sold to P. Harrison in March 1932 for £25. (This was the hull only). 1934 owner Braithwaite. 1939 hulk at Whitewall Creek, 1946. Broken up.

W.H.F. Official No. 56803
Her registered tonnage was 35 reg. tons and she carried 75 tons of cement. She was built at Millwall in 1867 and was owned by Robins & Co. of Northfleet in 1894, then sold to the Lighterage Company of London then the A.P.C.M. She was registered in Rochester. Still afloat as late as 1947. No known trace.

William Official No. 58467
Her registered tonnage was 41 reg. tons and she carried 80 tons of cement. She was built at Frindsbury by Geo. Curel in 1868 and

owned by The Burham Brick Lime & Cement Co., then A.P.C.M. Her dimensions were 73′7″ x 15′6″ x 4′9″. She was registered in Rochester. Sold to The Upnor Shipbreaking Co. in 1923 for £15. No known trace.

William Official No. 47946
Her registered tonnage was 39 reg. tons and she carried 80 tons of cement. She was built at Rochester in 1864 and owned by Henry Peters of Burham, then A.P.C.M. Her dimensions were 73′0″ x 14′5″ x 4′9″. She was registered in Rochester. Was sold to Viney Blaker in 1915 for £6. No known trace.

William & Sarah Official No. 18076
Her registered tonnage was 44 reg. tons and she carried 80 tons of cement. She was built at Frindsbury in 1853 and owned by Thomas Weekes, Trenchman Weekes of Halling, then the A.P.C.M. Her dimensions were 73′3″ x 14′4″ x 4′7″ – a canal or cut barge. Registered in Rochester. She was sold to Mr M. Hidges in 1922 for £5. No known trace.

W. Paxton Official No. 84377
Her registered tonnage was 52 reg. tons and she carried 130 tons of cement. She was built at Northfleet by The Lighterage Co. of London and owned by Robins & Co. of Northfleet, then the A.P.C.M. She was registered at Rochester. Was wrecked and lost on the coast of France on 23rd December 1918 at Treport Harbour entrance.

Winter Official No. 84409
Her registered tonnage was 44 reg. tons and she carried 90 tons of cement. She was built and owned by The Burham Brick, Lime & Cement Co. Aylesford, in 1881 and launched in August of that year. Her dimensions were 74′0″ x 14′6″ x 4′8″. P.L.A. No. 10516, registered in Rochester. She was sold to The Port of london Motor Boat Co. in May 1930 for £65. Derelict at Gravesend 1937.

Wouldham Court Official No. 109928
Her registered tonnage was 44 reg. tons and she carried 90 tons of cement. She was built at Rochester by M. Gill & Son in 1899 and owned by Henry Peters of Burham, then B.P.C.M. Her dimensions were 75′0″ x 14′11″ x 5′6″– was stumpie rigged. P.L.A. No. 14499, registered in Rochester. She was sold to The Tilbury Contracting Co. in May 1928 for £250. Later LRTC. Buried at Lower Upnor.

Wouldham Hall Official No. 109927
Her registered tonnage was 44 reg. tons and she carried 90 tons of cement. She was built at Rochester by M. Gill & Son in 1898 and owned by Henry Peters of Burham then the B.P.C.M. P.L.A. No. 14500, registered at Rochester. Was sold to Mr Moore in May 1928 for £150. No known trace.

SMEAD DEAN BARGES

The details and histories of the following seventy-one barges were the remainder of this once mighty Smeed Dean fleet. These craft were once sailing under the management of the 'Red Triangle' group up until 1932 when they became absorbed into the Combine. With this addition of these seventy-one barges to the fleet, it then brought the total up to a hundred and four barges. Of these additional craft no less than forty were built by Smeed Dean at their Murston bargeyard, and, in later years over thirty of these vessels had been re-built by them at Murston. Two had been converted into lighters and thirteen had been condemned.

Over the five years from 1930 to 1935 nearly all of the craft had been sold away and those that had been rebuilt gave very good service to their new owners.

Two were sold in 1931, eleven in 1932, twenty in 1933, sixteen in 1934 and thirteen in 1935. Eight were sunk or lost without trace.

The last barge working under sail of the huge Combine fleet was the ex *R.G.H.* which was re-built at Murston in 1928 and renamed the *Dunstable*.

This barge ran the Leigh-sand to the Murston brickfield up until 1928 and carried on in general trade till 1946. Not only was she the last of the Smeed Dean fleet to be working under sail, but also the last of the Combine fleet working under sail.

Ada Mary Official No. 94554
Her registered tonnage was 44 reg. tons and she carried 105 tons of cement. She was built at Murston in 1887 and registered in Rochester. Her dimensions were 76′7½″ x 18′0″ x 5′0″. She was rebuilt at Murston in 1915 and converted to a lighter in November 1931. Sold to Leigh Building Supply. Went to Dunkirk June 1940 in tow. At work 1950.

Agnes Official No. 106518
Her registered tonnage was 36 reg. tons and she carried 75 tons of cement. She was built at Conyer by A.M. White in 1896. Her dimensions were 74′4″ x 14′6″ x 4′9″. She was registered in Rochester. She was condemned in 1930s and sold to a Mr J. Brown for £5 (hull only) in May 1931. No known trace.

Alan Dean Official No. 125664
Her registered tonnage was 55 reg. tons and she carried 130 tons of cement. She was built at Murston in 1908. Her dimensions were 80′0″ x 19′0″ x 6′0″. She was registered in London. Was a coasting barge and foundered off Langney Point Eastbourne June 1935.

Alpha Official No. 104936
Her registered tonnage was 41 reg. tons and she carried 90 tons of cement. She was built at Rochester in 1896. Original owner S. Payn (Faversham). Her dimensions were 76′0″ x 17′0″ x 5′0″. She was registered at Faversham. Was sold to J. Brown for £15. in January 1933. Not on the 1934 register. No known trace.

Amy Official No. 68509
Her registered tonnage was 44 reg. tons and she carried 100 tons of cement. She was built at Reading in 1873. Her dimensions were 78′0″ x 17′0″ x 5′0″. She was registered in London. Coasting in 1908–11. Was condemned and sold to McDermott for £8.10s. in August 1932. Was a lighter 1931. Derelict 1940.

Argosy Official No. 78518
Her registered tonnage was 42 reg. tons and she carried 75 tons of cement. She was built at Sittingbourne in 1878. Her dimensions were 72′0″ x 14′0″ x 4′9½″. She was registered in Rochester. Was sold to a Mr A.P. Finch for £45 in April 1934. Hulked – Wallersea Island, River Crouch.

Bessie Official No. 94559
Her registered tonnage was 45 reg. tons and she carried 100 tons of cement. She was built at Murston in 1888. Her dimensions were 76′0″ x 17′0″ x 5′0″. She was registered in Rochester. Was rebuilt at Murston in 1909. Was sold to T. Allsworth of Queenborough for £55. Was a lighter 1945. Sunk in collision 1946.

Buckland Official No. 109912
Her registered tonnage was 35 reg. tons and she carried 80 tons of cement. She was built at Frindsbury in 1875. Her dimensions were 74′0″ x 14′7½″ x 5′0″. Was sold to T. Knight for £40 in March 1934.

Burton Official No. 81867
Her registered tonnage was 45 reg. tons and she carried 100 tons of cement. She was built at Murston in 1880. Her dimensions were 75′0″ x 17′0″ x 5′0″. She was registered in Rochester. Was rebuilt at Murston in 1914. Was sold to Leigh Building Society for £50 in May 1935. No known trace.

Curlew Official No. 84395
Her registered tonnage was 44 reg. tons and she carried 100 tons to sea. She was built at Milton in 1881. Was registered in Rochester. Was condemned and sold for £4 for breaking up in March 1932. Hulk at Borstal Yacht Basin 1937.

Derby Official No. 79866
Her registered tonnage was 32 reg. tons and she carried 80 tons of cement. She was built at Murston in 1878. Her dimensions were

73′0″ x 14′0″ x 5′0″ – a canal barge. She was registered in Rochester. Was rebuilt at Murston in 1910. She was sold to Bridge & Co. for £70 in March 1934. 1938 sailing under Leigh Building Supply.

Diligent Official No. 80521
Her registered tonnage was 36 reg. tons and she carried 90 tons of cement. She was built at Sittingbourne in 1880. Her dimensions were 73′0″ x 14′0″ x 5′0″ – a canal barge. She was registered at Faversham. Was sold to R.R. Race for £4 in March 1935. Derelict Leigh Creek 1957.

Ex-Smeed Dean's *Dunstable* in the 1937 barge match.
Note that she is now under the A.P.C.M.

Dunstable Official No. 98813 (ex. 'R.G.H.")
Her registered tonnage was 42 reg. tons and she carried 105 tons of cement. She was built at Milton in 1891 as the R.G.H. Her dimensions were 81′0″ x 19′0″ x 4′9½″. She was registered in Rochester. Was rebuilt at Murston in 1928. Was laid up in April 1946 and broken up in Northfleet Creek after 1951.

Easthall Official No. 99930
Her registered tonnage was 43 reg. tons and she carried 85 tons of cement. She was built at Sittingbourne in 1893. Her dimensions were 76′0″ x 16′0″ x 4′1″. Was registered at Rochester. This vessel was condemned after working as a clay barge. Sold to R. Evenden for £2 in October 1933. Derelict below Kingsferry Bridge 1934.

Ebenezer Official No. 80508
Her registered tonnage was 40 reg. tons and she carried 75 tons of

cement. She was built at Sittingbourne in 1879. Her dimensions were 75'0" x 17'0" x 5'0". Was registered at Faversham. She was condemned and sold to Mr E.L. Cox for 10s.0d. Broken up without trace.

Elsie Official No. 106513
Her registered tonnage was 40 reg. tons and she carried 90 tons of cement. She was built at Murston in 1896. Her dimensions were 75'0" x 16'0" x 4'0". Was registered in Rochester. She was rebuilt at Murston in 1925. Was sold to Sir H. Pallen in August 1933 as a barge yacht. No known trace.

Esther Official No. 112721
Her registered tonnage was 50 reg. tons and she carried 115 tons of cement. She was built at Murston in 1900. Her dimensions were 78'0" x 18'0" x 5'6". She was registered in London. She was sold to Mr Schmid for £70 in March 1933. 1945 to Sheppey Motor Barge Co. Finished work 1948. Fell to pieces at Ramsgate 1951.

Eureka Official No. 93980
Her registered tonnage was 45 reg. tons and she carried 100 tons of cement. She was built at Maldon in 1891. Her dimensions were 80'0" x 18'0" x 5'0". She was registered in Colchester. Was sold to W.G. Arnott for £125 in November 1934. Barge yacht. Broken up at Greenwich.

Favourite Official No. 11233
Her registered tonnage was 45 reg. tons and she carried 115 tons of cement. She was built at Sittingbourne in 1803. Her dimensions were 74'0" x 20'0" x 5'0". She was registered in Rochester. Was rebuilt in 1898 and 1926. She was sold to a Mr S.B. Shuffrey for £95 in April 1933. Was broken up at Chiswick.

Fred Official No. 84385
Her registered tonnage was 44 reg. tons and she carried 85 tons of cement. She was built at Murston in 1881. Her dimensions were 76'0" x 16'0" x 4'0". She was registered at Rochester. Was rebuilt at Murston in 1928. Sold to a Mr Higgs for £65 in March 1933. A lighter in London.

Gannet Official No. 95493
Her registered tonnage was 50 reg. tons and she carried 115 tons of cement. She was built at Limehouse in 1888. Her dimensions were 79'0" x 18'0" x 6'0". Was registered in London. Was rebuilt at Murston in 1927. Sold to Wakeley Brothers of Rainham Kent for £125 in March 1933. Foundered 9.3.1939 near W. Buxey Buoy on passage from Maidstone to Crouche with ragstone.

Garfield Official No. 84440
Her registered tonnage was 38 reg. tons and she carried 80 tons of

cement. She was built at Murston in 1882. Her dimensions were 74'0" x 14'0" x 5'0". Registered in Rochester. She was rebuilt at Murston in 1920. Was an Elmley Moorings barge. Derelict at Robins Creek Northfleet 1951.

George Official No. 81851
Her registered tonnage was 43 reg. tons and she carried 90 tons of cement. She was built at Murston in 1879. Her dimensions were 76'0" x 17'0" x 4'0". Registered in Rochester. Was rebuilt at Murston in 1903. Was sold at A.G. Wood for £25 in March 1935. Barge yacht. No known trace.

Georgiana Official No. 84416
Her registered tonnage was 46 reg. tons and she carried 100 tons of cement. She was built at Murston in 1881. Her dimensions were 76'0" x 16'0" x 4'0". Was registered in Rochester. She was sold (hull only) for £5 to R. Evenden in March 1934. (Was named after Mr George Smeed's eldest daughter). Derelict at Kings Ferry Bridge 1936.

George Smeed Official No. 84430
Her registered tonnage was 59 reg. tons and she carried 150 tons of cement. She was built at Murston in 1882. Her dimensions were 80'0" x 20'0" x 6'4". Registered in Rochester. Was rebuilt at Murston in 1922. Was a coasting barge and was sold to Francis & Gilders of Colchester on 20th November 1934. A lighter at Heybridge, then a barge yacht.

Gladstone Official No. 77628
Her registered tonnage was 41 reg. tons and she carried 85 tons of cement. She was built at Murston in 1877. Registered in Rochester. This vessel was condemned. Her hulk lies in The Swale, opposite Milton Creek.

Gordon Official No. 90971
Her registered tonnage was 44 reg. tons and she carried 100 tons of cement. She was built at Murston in 1884. Her dimensions were 76'0" x 17'0" x 5'0". Registered in Rochester. Was rebuilt at Murston in 1908. She was sold to W.H. Theobald for £50 in March 1935. Active till 1940. No known trace.

Gore Court Official No. 8443
Her registered tonnage was 45 reg. tons and she carried 90 tons of cement. She was built at Murston in 1882. Her dimensions were 76'0" x 17'0" x 5'1". Registered in Rochester. She was rebuilt in 1903 at Murston. Was sold to T. Schmid of Queenborough for £50. in March 1935. A motor barge in 1946. Broken up at Conyer Creek. Had the name *Vurtudort*.

Grace Official No. 98791
Her registered tonnage was 44 reg. tons and she carried 110 tons of cement. She was built at Murston in 1890. Her dimensions were 76′0″ x 18′0″ x 5′0″. Registered in Rochester. She was a converted lighter for one year, then sold to W.H. Theobald for £75. in August 1935. Sailing in 1938. Houseboat 1947.

Graham Official No. 108507
Her registered tonnage was 37 reg. tons and she carried 85 tons of cement. She was built at Murston in 1897 and was registered in Rochester. She was condemned after working as a clay barge. Was sold to Mr McDermott for £8.10s. in August 1932. No known trace.

Gertrude May Official No. 99932
Her registered tonnage was 72 reg. tons and she carried 150 tons of cement. She was built at Sittingbourne in 1893. Her dimensions were 85′0″ x 20′0″ x 6′0″. She was registered in Rochester. Was rebuilt at Murston in 1921. Was a coasting barge and was sold to Francis Gilders for £520 in January 1934. Mined on 27.12.42 with a cargo of flour.

Ex-Smeed Dean's *Hambrook* built as the *Morley* circa 1933.

Hambrook Official No. 87212
Her registered tonnage was 38 reg. tons and she carried 115 tons of cement. After rebuilding was 49 reg. tons. She was built in Rochester and registered in 1877 as the *Morley* . Her dimensions were

76′0″ x 18′9″ x 6′0″. Original *Hambrook* was broken up after sinking in collision 9.2.1921. Was rebuilt at Murston in 1923 and re-named. She was a coasting barge. Was sold to A. Sully for £200 in October 1934.

Harry Official No. 110002
Her registered tonnage was 46 reg. tons and she carried 110 tons of cement. She was built at Murston in 1898. Her dimensions were 83′0″ x 18′0″ x 6′0″. Registered in London. Was sunk in Ridham Dock, Kemsley, nr. Sittingbourne in May 1932 (ballast laden). She was sold to R. Evemden for £25 in May 1932, resold as a yacht club hulk. Her hulk lies on the River Crouch side of Wallersea Island.

Hilda Official No. 104054
Her registered tonnage was 57 reg. tons and she carried 130 tons of cement. She was built and registered in Ipswich in 1895. SD purchased her after sinking. Her dimensions were 83′0″ x 18′0″ x 6′0″. She was sold to C.H. Barclay for £100 in September 1933. No known trace.

Hilsted Official No. 81171
Her registered tonnage was 53 reg. tons and she carried 100 tons of cement. She was built at Ipswich in 1880 as the *Lydia*. Capsized off Dover 10.12.1896 and renamed by SD after repair. Her dimensions were 80′0″ x 18′0″ x 6′0″. She was registered in London. Was condemned and sold to R. Evenden for £25 in May 1933. A leaky old craft only fit to carry coke in her latter days.

Jane Mead Official No. 90981
Her registered tonnage was 42 reg. tons and she carried 90 tons of cement. She was built at Murston in 1886. Her dimensions were 76′0″ x 17′0″ x 5′0″. She was registered in Rochester. Was rebuilt at Murston in 1903. She was sold to F.H. Wells for £55 in April 1935. Was a barge yacht. No known trace.

Jessie Official No. 94564
Her registered tonnage was 42 reg. tons and she carried 105 tons of cement. She was built at Murston in 1888. Her dimensions were 76′0″ x 18′0″ x 5′0″. She was registered in Rochester. Was rebuilt at Murston in 1916. She was sold to T. Allsworth of Queenborough, for £80 in September, 1935. Stranded at Queenborough 1943. Derelict at Murston 1950.

Joe Official No. 109963
Her registered tonnage was 49 reg. tons and she carried 110 tons of cement. She was built at Murston in 1898. Her dimensions were 78′0″ x 18′0″ x 5′5″. She was registered in Rochester. Sold to Mr R.T. Turner in October 1933 for £100. Trading to Southend prior to 1939. House boat 1939.

Leslie Official No. 104303
Her registered tonnage was 43 reg. tons and she carried 90 tons of cement. She was built at Murston in 1894. Her dimensions were 76′0″ x 16′0″ x 4′6″. Was registered at Rochester. She was rebuilt at Murston in 1926. Sold to Geo. Andrews of Sittingbourne for £150 on 23rd July 1933. Barge yacht at Allington 1947-76.

Levitt Official No. 98811
Her registered tonnage was 44 reg. tons and she carried 100 tons of cement. She was built at murston in 1891 and registered in Rochester. Her dimensions were 76′0″ x 17′0″ x 5′0″. She was rebuilt at Murston in 1915. Was sold to A.G. Wood of Sittingbourne in October 1935 for £65. Resold as a barge yacht. Burnt-out St. Albans Head circa 1936.

Livingstone Official No. 81883
Her registered tonnage was 44 reg. tons and she carried 85 tons of cement. She was built at Murston in 1880. Her dimensions were 76′0″ x 16′0″ x 4′0″. She was registered in Rochester. Was condemned after working as a clay barge. Sold to Mr A. Smith for £3 in June 1933. Derelict Kings Ferry Bridge 1936.

Lizzie Official No. 99918
Her registered tonnage was 52 reg. tons and she carried 115 tons of cement. She was built at Murston in 1892. Her dimensions were 79′0″ x 18′0″ x 5′2″. She was registered at Rochester. Sold to Mr G.H. Williams for £50 in January 1934. No known trace.

Longfield Official No. 76589
Her registered tonnage was 47 reg. tons and she carried 70 tons of cement. She was built and registered in Rochester in 1876. Her dimensions were 73′0″ x 14′0″ x 4′0″. She was condemned and sold to Mr McDermott for £8.10s. in 1932. Derelict in Damhead Creek 1940.

Lowe Official No. 79851
Her registered tonnage was 40 reg. tons and she carried 80 tons of cement. She was built at Murston in 1878. Her dimensions were 73′0″ x 14′4″ x 4′0″. She was registered in Rochester. Was sold to John Mowlem & Co. for £60 in September 1933. No known trace.

Maid of Connaught Official No. 112612
Her registered tonnage was 68 reg. tons and she carried 150 tons of cement. She was built at East Greenwich in 1899 as the *Monarch*. Her dimensions were 85′0″ x 20′7″ x 5′9″ (must have been a big barge). She was registered in Rochester. Was rebuilt at Murston in 1929 and sold to W. Theobald for £250 in May 1934. A coasting barge, converted to a motor barge 1946.

Maid of Munster Official No. 109918
Her registered tonnage was 42 reg. tons and she carried 105 tons of

cement. She was built at Sittingbourne in 1898 as the *Bexhill.* Her dimensions were 76'0" x 17'0" x 5'1½". She was registered in Rochester. Was rebuilt at Murston in 1928. Sold to the Leigh Building Society for £175 in September 1933. Converted to a motor barge 1946. Sold to Wakeley Bros. 1952.

Maria Official No.
Her registered tonnage was 36 reg. tons and she carried 80 tons of cement. She was built at Murston in 1891 by Smead Deans. Her dimensions were 74'0" x 14'0" x 5'0". Registered in Rochester. Was rebuilt at Murston in 1912. Sunk in collision with S.S. *Ariel* on 27th October 1931. Sold to Mr A. Smith for £3 in June 1933. Had a Ford 'T' Engine fitted. A canal barge. Derelict Shepherds Creek 1950.

Marjorie Official No. 113688
Her registered tonnage was 56 reg. tons and she carried 150 tons of cement. She was built and registered at Rochester in 1900. Her dimensions were 82'0" x 19'0" x 5'7". She was sold to Mr A.G. Jones for £65 in June 1933.

Martin Luther Official No.
Her registered tonnage was 43 reg. tons and she carried 85 tons of cement. She was built at Murston in 1884 and registered in Rochester. Was sold to A.G. Wood of Sittingbourne for £25 in June 1935. Yacht barge 1936. No known trace.

Mary Ann Official No.
Her registered tonnage was 46 reg. tons and she carried 110 tons of cement. She was built at Murston in 1900 by Smeed Deans. Her dimensions were 78'0" x 18'0" x 5'0". Registered in Rochester. Sold to Theobald & Bridger for £320 in December 1933. Coverted to a motor barge. No known trace.

McKinley Official No. 114809
Her registered tonnage was 50 reg. tons and she carried 115 tons of cement. She was built at Sittingbourne in 1901, was a coasting barge and owned by Wills & Packham. Her dimensions were 80'0" x 18'0" x 5'0". Registered in London. She was sold to Tester Bros. for £200 in October 1934. A lighter 1939. No known trace.

Murston Official No. 67106
Her registered tonnage was 37 reg. tons and she carried 85 tons of cement. She was built at Sittingbourne in 1885 as the *Levetta*. Her dimensions were 77'0" x 16'2½" x 5'2½". She was rebuilt at Murston in 1905. Sold to W.H. Millway for £40 in July 1934. No known trace.

Myrtle Official No. 81864
Her registered tonnage was 38 reg. tons and she carried 80 tons of cement. She was built at Milton in 1880, for G. Trowell. Later owned by Cremer. Her dimensions were 74'0" x 16'0" x 5'0". Registered in

Rochester. She was sold to Mr H. Compton for £50 in February 1934. No known trace.

Smeed Dean's *Nicholas* in the 1929 Medway barge match

Nicholas Official No. 112776
Her registered tonnage was 47 reg. tons and she carried 130 tons of cement. She was built at Faversham in 1900 by A. White. Her dimensions were 83′0″ x 19′0″ x 5′7″. Registered in London. Her hull was sold to The Thames Transport Co. for £75 in June 1932 as a lighter.

Oxygen Official No. 104329
Her registered tonnage was 69 reg. tons and she carried 160 tons of cement. She was built and registered in Rochester in 1895 by A. Gill. Her dimensions were 84′0″ x 20′0″ x 6′8″. She was rebuilt at Murston in 1926. Sold to A. Sully for £540 in April 1934. She was a coasting barge. Auxiliary engine fitted in 1948 from the *Arcades.*

Persevere Official No. 94572
Her registered tonnage was 58 reg. tons and she carried 130 tons of cement. She was built at Murston in 1889 by Smeed Deans. Her dimensions were 80′8″ x 19′0″ x 6′2½″. She was registered in Rochester Sold to Theobald-Bridge for £150 in February 1934. She was a coasting barge. Converted to a motor barge 1946.

Plimsoll Official No. 78537
Her registered tonnage was 41 reg. tons and she carried 80 tons of cement. She was built at Murston in 1878 by Smeed Deans. Her

dimensions were 72′0″ x 14′0″ x 4′0″. She was registered in Rochester. Was sold to Calvert Bros. for £16.10s. in July 1934. A canal or cut barge.

Providence Official No. 76596
Her registered tonnage was 36 reg. tons and she carried 80 tons of cement. She was built at Sittingbourne in 1877, owners 1889 W. Cremer Her dimensions were 73′0″ x 15′2″ x 5′0″. She was registered in Rochester. Was condemned and sold to Mr McDermott for £8.10s. in August 1932. No known trace.

Russell Official No. 79895
Her registered tonnage was 44 reg. tons and she carried 105 tons of cement. She was built at Murston by Smead Dean in 1879. Her dimensions were 76′0″ x 18′0″ x 5′0″. Registered in Rochester. She was rebuilt at Murston in 1920. Sold to W.H. Theobald for £75 in July 1935. Yacht barge 1950. House boat Strood 1960.

Sam Official No. 104326
Her registered tonnage was 39 reg. tons and she carried 85 tons of cement. She was built at Murston in 1895 by Smead Dean. She was registered in Rochester. Was condemned and sold to Mr McDermott for £8.10s. in August 1932.

Spurgeon Official No. 87219
Her registered tonnage was 48 reg. tons and she carried 110 tons of cement. She was built at Murston by Smead Dean in 1883. Her dimensions were 76′0″ x 17′0″ x 5′0″. She was registered in Rochester. Was rebuilt at Murston in 1924. Sold to W.H. Theobald for £80 in October 1935. Was at Dunkirk 1940. Sunk at Allington Locks, River Medway after being a houseboat, 1960. Broken up at Allington.

Valders Official No. 97722
Her registered tonnage was 45 reg. tons and she carried 90 tons of cement. She was built at Sittingbourne in 1890. Her dimensions were 76′0″ x 17′0″ x 5′1½″. She was registered at Rochester. She was sold to Mr H.G. Lancaster for £115 in September 1932. Sails removed Engine fitted circa 1934. No known trace.

Vavasour Official No. 108205
Her registered tonnage was 64 reg. tons and she carried 160 tons of cement. She was built at Blackwall in 1897. Her dimensions were 82′0″ x 20′0″ x 5′6″. A very slow barge. She was registered in Rochester. Was converted to a lighter in November 1931. Crane hulk in 1940. No known trace.

V.C. Official No. 109961
Her registered tonnage was 50 reg. tons and she carried 115 tons of

cement. She was built at Murston by Smead Dean in 1898 as the *Donald*. She was rebuilt in 1920 and renamed *V.C.* after Donald Smeed won the V.C. 1914. Her dimensions were 79′0″ x 18′3″ x 5′0″. She was sold to A.G. Wood for £7.10s. in October 1935. House boat at Adelaide Dock, Murston 1938.

Victoria Official No. 108233
Her registered tonnage was 54 reg. tons and she carried 135 tons of cement. She was built at Sittingbourne in 1897. Her dimensions were 85′0″ x 18′0″ x 5′6″. She was registered in London. Was sold to Mr E.J. Butcher for £400 in November 1933. Yacht barge 1950. House-boat on River Stour.

Ex Smeed Dean *Vincent* circa 1933.

Vincent Official No. 79879
Her registered tonnage was 35 reg. tons and she carried 80 tons of cement. She was built at Murston. Registered in Rochester. Her dimensions were 73′0″ x 14′0″ x 5′0″. She was rebuilt at Murston in 1902. She was a moorings barge at Hollicks & Co. 'Morden Wharf', East Greenwich. A canal or cut barge. No known trace.

Wave Official No. 45237
Her registered tonnage was 37 reg. tons and she carried 80 tons of cement. She was built and registered at Faversham in 1862. She was condemned and sold to Mr McDermott for £8.10s. in August 1932. Derelict at Damhead Creek 1940.

Whitehall Official No. 84399
Her registered tonnage was 37 reg. tons and she carried 80 tons of

cement. She was built at Murston in 1881 by Smead Dean and was registered at Rochester. Her dimensions were 73′0″ x 14′0″ x 5′0″. She was rebuilt at Murston in 1913. A canal or cut barge. Derelict at Rayfields Yard, Northfleet 1947.

Winnie Official No. 81896
Her registered tonnage was 45 reg. tons and she carried 85 tons of cement. She was built at Murston in 1880 by Smead Deans. She was registered at Rochester. Was condemned after working as a clay barge. She was sold to Mr A.S. Smith for £3 in June 1933. Broken up at Rushenden, Sheppey, 1940.

Smeed Dean's *Youngarth* at Murston 1912.

Youngarth Official No. 135312
Her registered tonnage was 63 reg. tons and she carried 160 tons of cement. She was built at Murston in 1913 by Smead Deans and registered in London. Her dimensions were 83′0″ x 20′0″ x 6′0″. She was converted to a motor barge, fitted with a 30hp Atlantic engine. 1951 collar barge at E. Greenwich. Broken up at Upnor in September 1962. She was a coasting barge.

Youngjack Official No. 81884
Her registered tonnage was 44 reg. tons and she carried 85 tons of cement. She was built at Murston in 1880 by Smead Deans. Her dimensions were 76′0″ x 16′5½″ x 4′8½″. She was registered at Rochester. Was condemned after working as a clay barge. She was sold to Mr McDermott for £8.10s. in August 1932.

THE THAMES PORTLAND CEMENT WORKS
GIBBS & Co
TRADE MARK
DIAMOND BRAND
REGISTERED
GRAYS ESSEX

Wm LEE. SON & Co
PORTLAND
TRADE MARK.
CEMENT.
MADE IN ENGLAND
LONDON.

LONDON PORTLAND CEMENT Co
SEMPER IDEM.
LONDON NORTHFLEET

THE BRITISH PORTLAND CEMENT MANUFACTURERS LIMITED
BEST ENGLISH
TRADE MARK
FALCON BRAND
PORTLAND CEMENT
-LONDON-

PORTLAND CEMENT
Peters Bros
PETERS
LONDON
WOULDHAM HALL & BURHAM WORKS, KENT.

THE BRITISH PORTLAND CEMENT MANUFACTURERS LIMITED
GOLIATH BRAND
PORTLAND CEMENT
LONDON.

BURHAM BRICK LIME AND CEMENT Co
BURHAM
TRADE
PORTLAND CEMENT
MARK
WORKS
-LONDON-

PORTLAND
BLUE CIRCLE
CEMENT

J. B. WHITE & BROTHERS
LONDON

GLOBE PORTLAND CEMENT & WHITING Co
PORTLAND CEMENT
GLOBE
ESTABLISHED 1869
LONDON

GILLINGHAM PORTLAND CEMENT COMPANY
LONDON
PORTLAND CEMENT
REGISTERED
ENGLAND

THE ASSOCIATED PORTLAND CEMENT MANUFACTURERS LIMITED
BRIDGE BRAND
ENGLISH PORTLAND CEMENT
LONDON

THE WOULDHAM CEMENT COMPANY LIMITED
LONDON
LION WORKS
WEST THURROCK, ESSEX.
ENGLAND

THE WEST KENT PORTLAND CEMENT CO.
INVICTA BRAND
LONDON

BEST ENGLISH PORTLAND CEMENT
MARTIN EARLE & COMPANY LIMD
ROCHESTER, KENT.
LONDON OFFICES.
159, QUEEN VICTORIA STREET, E.C.